MIX
Papier aus verantwortungsvollen Quellen
Paper from responsible sources
FSC® C105338

Ripandeep Singh

Characterization of microstructure and mechanical properties of AL6063 using FSP Multipass

Anchor Academic
Publishing

Singh, Ripandeep: Characterization of microstructure and mechanical properties of
AL6063 using FSP Multipass, Hamburg, Anchor Academic Publishing 2018

Buch-ISBN: 978-3-96067-211-1
PDF-eBook-ISBN: 978-3-96067-711-6
Druck/Herstellung: Anchor Academic Publishing, Hamburg, 2018

Bibliografische Information der Deutschen Nationalbibliothek:
Die Deutsche Nationalbibliothek verzeichnet diese Publikation in der Deutschen
Nationalbibliografie; detaillierte bibliografische Daten sind im Internet über
http://dnb.d-nb.de abrufbar.

Bibliographical Information of the German National Library:
The German National Library lists this publication in the German National Bibliography.
Detailed bibliographic data can be found at: http://dnb.d-nb.de

All rights reserved. This publication may not be reproduced, stored in a retrieval system
or transmitted, in any form or by any means, electronic, mechanical, photocopying,
recording or otherwise, without the prior permission of the publishers.

Das Werk einschließlich aller seiner Teile ist urheberrechtlich geschützt. Jede Verwertung
außerhalb der Grenzen des Urheberrechtsgesetzes ist ohne Zustimmung des Verlages
unzulässig und strafbar. Dies gilt insbesondere für Vervielfältigungen, Übersetzungen,
Mikroverfilmungen und die Einspeicherung und Bearbeitung in elektronischen Systemen.

Die Wiedergabe von Gebrauchsnamen, Handelsnamen, Warenbezeichnungen usw. in
diesem Werk berechtigt auch ohne besondere Kennzeichnung nicht zu der Annahme,
dass solche Namen im Sinne der Warenzeichen- und Markenschutz-Gesetzgebung als frei
zu betrachten wären und daher von jedermann benutzt werden dürften.

Die Informationen in diesem Werk wurden mit Sorgfalt erarbeitet. Dennoch können
Fehler nicht vollständig ausgeschlossen werden und die Diplomica Verlag GmbH, die
Autoren oder Übersetzer übernehmen keine juristische Verantwortung oder irgendeine
Haftung für evtl. verbliebene fehlerhafte Angaben und deren Folgen.

Alle Rechte vorbehalten

© Anchor Academic Publishing, Imprint der Diplomica Verlag GmbH
Hermannstal 119k, 22119 Hamburg
http://www.diplomica-verlag.de, Hamburg 2018
Printed in Germany

ABSTRACT

Need for low weight and high performance structural materials have revolutionized the technology and had led to the emergence of new processes and methodologies. Frictions stir processing (FSP), based on principle of friction stir welding, is an emerging solid state metal working process. This technique causes intense plastic deformation and high strain rates in the processed material resulting in precise control of the microstructure through material mixing and densification. FSP process has been successfully used for achieving significant grain refinement and enhancement of surface properties.

Present work is focused on the study of behavior of Aluminium cast alloy (Al-6063) with processed by friction stir processing technique. Samples of FSPed aluminium were examined and their microstructures, microhardness, Rockwell hardnesss, impact strength were studied and compared with base metal Al-6063.

Hardness tester is employed to evaluate the interfacial bonding between the particles and matrix by indenting the hardness with the constant load and constant time. Impact test is employed to know the Impact Strength of samples against the Impact of Hammer.

CONTENTS

Chapter 1 Introduction 1

    1.1 Friction Stir Processing 1
    1.2 FSP Parameters 13
        1.2.1 Tool Geometry 13
        1.2.2 Processing Parameters 14
        1.2.3 Tool Tilt Angle 14
    1.3 Process Modelling 15
        1.3.1 Metal Flow 15
        1.3.2 Temperature Distribution 15
    1.4 Microstructural evolution 16
        1.4.1 Nugget Zone 16
        1.4.2 Thermo-mechanically affected zone 18
        1.4.3 Heat-affected zone 19
    1.5 Applications of FSP 19
        1.5.1 Superplasticity 19
        1.5.2 Surface composites 19
        1.5.3 Microstructural Modification 20
    1.6 Need For Present Study 21
    1.7 Chapter Scheme 22

Chapter 2 Literature Review 23

    2.1 Review of related literature 23
    2.2 Summary of related literature 36
    2.3 Research gaps 37

2.4 Objective of study	37
2.5 Methodology	37

Chapter 3 Experimentation — 38

3.1 Sample Preparation	38
3.1.1 Application of AL6063	39
3.1.2 Fabrication of AL6063	39
3.1.3 Supplied Forms	39
3.1.4 Properties of Aluminium AL6063	40
3.1.5 Welding of 6063 Aluminium	41
3.2 FSP Setup	41
3.2.1 Tool Geometry	42
3.2.2 Tool Specifications	42
3.2.3 CNC Vertical milling machine	42
3.2.4 Fixture	44
3.2.5 Machining Parameters	45
3.2.6 Different results after experimentation	45
3.3 Characterization and Testing	46
3.3.1 Microstructural Examination	46
3.3.2 Microhardness Measurement	48
3.3.3 Izod Impact test	49
3.3.4 Rockwell hardness test	51

Chapter 4 Results and Discussions — 53

4.1 Evaluation of Mechanical Properties	53
4.1.1 Microhardness Evaluation	53
4.1.2 Rockwell hardness test	55
4.1.3 Optical Microscopy	56
4.1.4 Izod Impact result	57

Chapter 5 Conclusions and Recommendations — 59

5.1 Scope for Future Work	59

References — 60

LIST OF FIGURES

Figure no.	Title.	Page no.
1.1	A FSP Process	1
1.2	A FSP Machine	2
1.3	A Robotic FSP Machine	3
1.4	FSP Tool Movement	5
1.5	Impression Formed by Tool on Work piece	6
1.6	Travelling and Rotational Movement of Pin	7
1.7	The Different Tool Profiles	7
1.8	Experimental Setup	10
1.9	FSP Setup	12
1.10	FSP Laminate	13
1.11	Schematic Drawing of FSP Tool	14
1.12	Various Microstructural Zones in FSP	16
1.13	Effect of Processing Parameters on Nugget Shape	17
1.14	Microstructure of Thermo mechanically Effected Zone in FSP	18
3.1	Aluminium Alloy AL6063 Sample	40
3.2	FSP Tool	41
3.3	CNC Vertical Milling Machine	42
3.4	Fixture used for Holding the Specimen	44
3.5	Shows the used Tool	45
3.6	Experimental Parameters	45
3.7	Sample Impressions After Conducting FSP	46
3.8	Inverted Optical Microscope	47
3.9	Mounted Specimen	47
3.10	Buffing Machine	48
3.11	Microhardness Tester	49
3.12	Impact Testing Machine	49
3.13	Different Results After Impact Test	50
3.14	Rockwell Test Indenter	51
3.15	Rockwell Hardness Tester	51
3.16	Hardness Tester Specifications	52
3.17	Hardness Tested Sample	52

4.1	Graph Showing Microhardness Results	54
4.2	Graph Showing Rockwell Hardness	55
4.3	Microstructure of Samples after different passes	56
4.4	Graph Comparison Between Impact Strength	58

LIST OF TABLES

Table no.	Title.	Page no.
1.1	A summary of grain size in nugget zone	18
2.1	Summary of related literature	36
3.1	Physical composition of Al6063	38
3.2	Tempers of Al6063	40
3.3	Physical properties of AL6063	40
3.4	Specification of tool	42
3.5	Specification of CNC machine	43

CHAPTER 1
INTRODUCTION

1.1 Friction Stir Processing

FSP is manufacturing the Technique used to modify the microstructure metals. In this process a rotating tool is penetrated in the work piece and moved in the transverse direction.

Further FSP technique is used for fabrication of surface composite on aluminium substrate and homogenization of powder metallurgy aluminium alloy, metal matrix composites, and the cast aluminium alloys. By this process material properties can be improved due to enhancement of grain structure. With this technique the material have shown good corrosion resistance, high strength and high fatigue resistance.

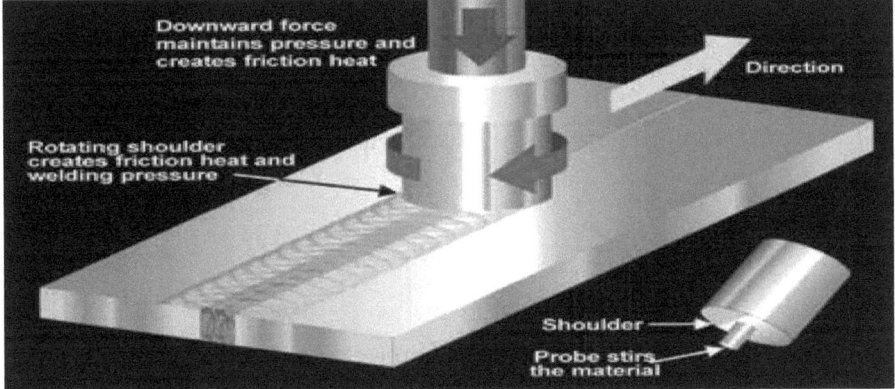

Fig 1.1 A FSP process (*http://en.wikipedia.org/wiki/Friction_stir_processing*)

Friction stir processing is a method of changing the properties of a metal through intense, localized deformation. This deformation is produced by forcibly inserting a non-consumable tool into the workpiece, and revolving the tool in a stirring motion as it is pushed laterally through the workpiece. The precursor of this technique, friction stir welding, is used to join multiple pieces of metal without creating the heat effected zone typical of fusion welding. When ideally implemented, this process mixes the material without changing the phase and creates a microstructure with fine, equixed grains. This homogeneous grain structure, separated by high-angle boundaries, allows some aluminium alloys to take on superplastic properties. Friction stir processing also enhances the tensile strength and fatigue strength of the metal. In tests with actively cooled magnesium-alloy work pieces, the microhardness was almost tripled in the area of the friction stir processed seam. (http://en.wikipedia.org/wiki/Friction stir processing)

It is widely used in aerospace industries and used in automobile industry for making body panels, for cycle frames and other components, these alloys are used for boat building and shipbuilding, It is used for architectural fabrication, particularly window frames, door frames, roofs, and sign frames. Furthermore, the FSP technique has been used for the fabrication of a surface composite on aluminium substrate and the homogenization of powder metallurgy (PM) aluminium alloys, metal matrix composites, and cast aluminium alloys. Compared to other metalworking techniques, FSP has distinct advantage

Fig 1.2 A FSP Machine – 3 axis (Mishra *et al.,* 2005)

Friction Stir Welding (FSW) is being targeted by modern industries for structurally demanding applications providing high-performance benefits. FSW has been shown to strongly decrease severe distortion and residual stresses compared to the traditional welding processes. The Frictioned zone consists of a weld nugget, a thermo-mechanically affected zone and a heat affected zone. The process results in obtaining a very fine and equiaxed grain structure in the weld nugget causing a higher mechanical strength and ductility. Jata and Semiatin showed that the microstructure in the weld nugget zone evolves through a continuous dynamical recrystallization process, the strong grain refinement produced by the process leads the microstructure to the fine dimensions offering the possibility to exhibit superplastic properties.

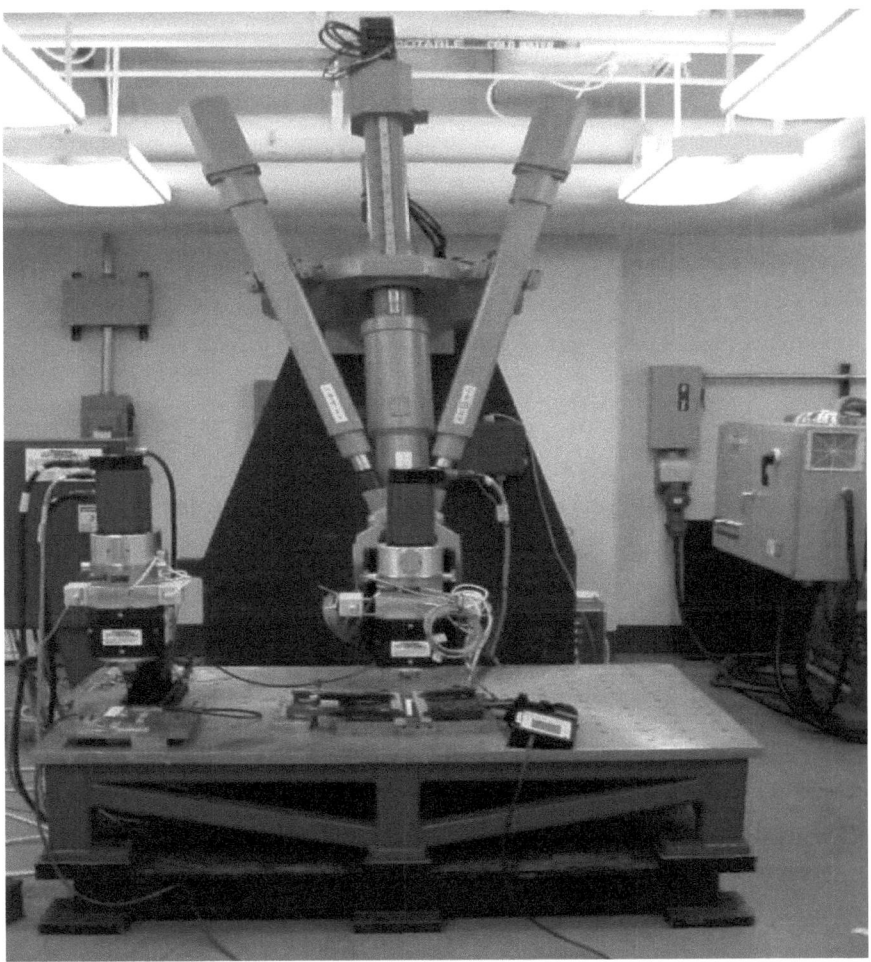

Fig 1.3 A Robotic FSP Machine – 6 axis (Mishra *et al.*, 2005)

Such technology requires a thorough understanding of the process and consequent mechanical properties of the heavily deformed material in order to be used in the production of components for aerospace applications and for this reason detailed research and qualification work is required. In the FSP, a rotating tool with a specially designed rotating probe travels down the surfaces of metal plates, and produces a highly plastically deformed zone through the associated stirring action. The localized thermo-mechanical effected zone is produced by friction between the tool shoulder and the plate top surface, as well as plastic deformation of the material in contact with the tool. The probe is typically slightly shorter than the thickness of the work piece and its diameter is typically the thickness of the work piece. The FSP process is a solid state process and therefore a solidification structure is absent and the problem related to the presence of brittle inter-dendritic and eutectic phases is eliminated. In addition the strong grain refinement and the possibility to obtain a uniform microstructure has lead researchers to investigate the possibility to employ such a process to increase the superplastic properties of some aluminium alloys. In FSP the work piece does not reach the melting point and the mechanical properties of the material are much higher compared to the traditional techniques, in fact, the undesirable microstructure resulting from melting and re-solidification, characterized by low mechanical properties, is absent leading to improved mechanical properties such as ductility and strength in some alloys. (Cavaliere *et al.*, 2005)

Friction stir processing (FSP) is a unique technique for refining and modifying the microstructure of materials. In many aspects, the basic principles of FSP are the same as those of FSW. Besides, FSP has been successfully used for fabricating the surface composites, for refining the surface microstructures of various materials as well as for synthesizing the composite and intermetallic compounds.

Friction stir processing is based on the friction stir welding (FSW) technique which was invented by The Welding Institute (TWI) in 1991. On observing the advantages associated with FSW, mainly grain refinement, the phenomenon has been extended to processing of commercial alloys. Friction stir processing (FSP) is a solid-state process in which a specially designed rotating cylindrical tool, consisting of a pin and a shoulder, is plunged into the sheet. The tool is then traversed in the desired direction. The rubbing of the rotating shoulder generates heat which softens the material (below the melting temperature of the sheet) and with the mechanical stirring caused by the pin, the material within the processed zone undergoes intense plastic deformation yielding a dynamically recrystallized fine grain structure. Despite the large number of studies that are being conducted to advance FSP technology, the effects of FSP on various mechanical and microstructural properties are still in need for further investigations. In addition, correlations between FSP parameters, mechanical properties and microstructural characteristics are not yet well understood. Accurate correlations are needed for successful modeling and process optimization. Most of the work that has been done in the field of friction stir processing focuses on aluminium alloys. (Darras *et al.*, 2007)

Friction stir processing (FSP), developed based on the basic principles of friction stir welding (FSW), a solid-state joining process originally developed for aluminium alloys, is an emerging metal working technique that can provide localized modification and control of microstructures in near-surface layers of processed metallic components. The FSP causes intense plastic deformation, material mixing, and thermal exposure, resulting in significant microstructural refinement, densification, and homogeneity of the processed zone. The FSP technique has been successfully used for producing the fine-grained structure and surface composite, modifying the microstructure of materials, and synthesizing the composite and intermetallic compound in situ. In this review article, the current state of the understanding and development of FSP is addressed. (Ma *et al.*, 2008)

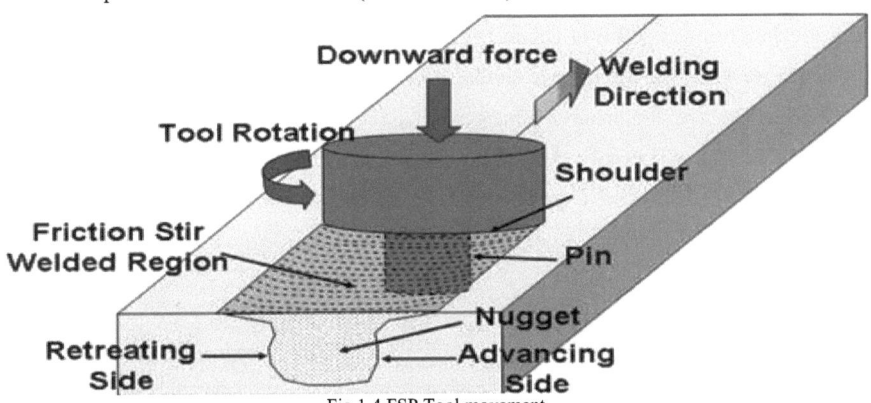

Fig 1.4 FSP Tool movement

Friction stir welding (FSW) is a relatively new solid state joining process. It requires no local melting in order to join pieces together. This joining technique is energy efficient, environment friendly, and versatile. The basic concept of FSW is remarkably simple. A non-consumable rotating tool with a specially designed pin and shoulder is inserted into the abutting edges of sheets or plates to be joined and traversed along the line of joint. The tool serves two primary functions: (a) heating of work piece, and (b) movement of material to produce the joint. The heating is accomplished by friction between the tool and the work piece and plastic deformation of work piece. During FSW process, the material undergoes intense plastic deformation at elevated temperature, resulting in generation of fine and equiaxed recrystallized grains. The fine microstructure in friction stir welds produces good mechanical properties. Recently friction stir processing (FSP) was developed by Mishra and Mahoney as a generic tool for microstructural modification based on the basic principles of FSW. In this case, a rotating tool is inserted in a monolithic work piece which provides frictional heating and mechanical mixing in the area covered by the tool. The large processing strain involved, resulted in microstructural refinement and homogenization. FSP has developed into a broad field covering microforming, casting modification, powder processing. Another area in which FSP shows a lot of promise is in the creation of superplastic materials. FSP creates a region called the "stir zone" (SZ) or "nugget", where the microstructural refinement occurs with equiaxed ultra-fine grains with high angle grain boundaries. The resultant microstructure in the nugget region can present the ideal conditions for superplasticity in some materials. (Ehab *et al.,* 2010)

FSP is a solid state processing based on friction stir welding (FSW). FSW is innovation of The Welding Institute (TWI) of UK in 1991. FSP is defined as a severe plastic deformation technique used in order to improve surface properties. Fine microstructure and absence of casting defects are the main advantages of FSP. During FSP, rotational tool (non-consumable) with a specially designed pin and shoulder is plunged into the plate and traversed in desired direction. (Akramifard *et al.,* 2014)

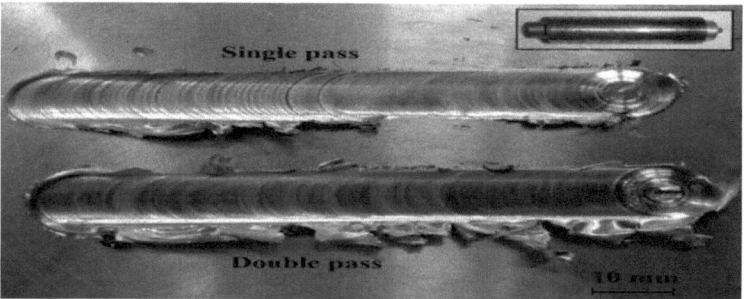

Fig 1.5 Impression formed on workpiece after FSP mutipass

Multipass FSP can be performed in two ways, continuously or after cooldown of first pass. multipass in which material was allowed to cool back down to room temperature and after that second pass was employed and second method is all subsequent passes were employed continuously without alloying the material to cool down.

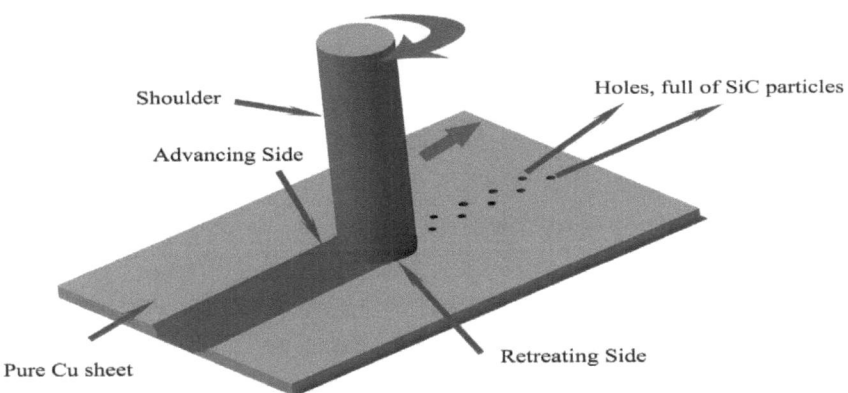

Fig 1.6 Travelling and rotational movement of pin and fabrication of Cu/Sic

Friction stir welding (FSW) technique was first invented by The Welding Institute (TWI), widely used in welding of aluminium, magnesium and copper alloys. This technique not only being limited to welding, friction stir processing (FSP) has found various applications: grain refinement of wrought and casting parts, super plasticity improvement, formation of intermetallics and composite fabrication. FSP offers a low energy consumption route to introduce reinforcing phases into the metal matrix and to form bulk composites. Grain refinement is also enhanced by dynamic recrystallization and grain boundary pinning during FSP.

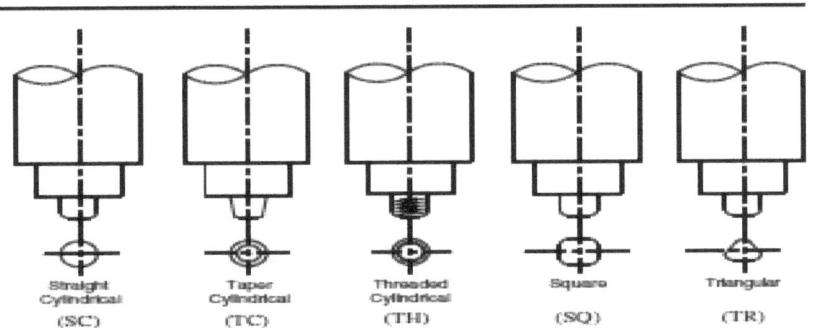

Fig 1.7 shows the different FSP tool profiles (Elangovan *et al.*, 2008)

Different types of tool profiles have been used for FSP process for different type of result to be obtained. Above Fig. 1.7 shows different tool profiles are Cylindrical, Taper cylindrical, threaded cylindrical, square, and Triangular. Friction stir processing (FSP) is a severe plastic deformation method that is used to produce bulk samples of fine-grained microstructure utilizing the same technique employed to join samples in friction stir welding (FSW). The thermo mechanical history of a sample produced through FSP affects the final grain size and therefore, the mechanical properties of the bulk materialThe effect of tool profile can be checked on the microstructure and various mechanical properties by using different testing equipments. The main effect in FSP is due to the heat generated which result in material deformation and microstructural changes. In this process 86% of the heat generated is from the shoulder, 11% from the sides of the bit and rest 3 % of heat is generated at the tip of the bit in FSP process. (Douglas *et al.*, 2007).

Friction stir processing (FSP) and friction stir welding (FSW) are allied technologies involving localized severe plastic deformation induced by the action of a non-consumable tool on a deformable material. In both FSP and FSW, the tool generally consists of a cylindrical shoulder portion with a projecting, concentric, and smaller-diameter pin. The tool is rotated while the pin is forced into the surface of the workpiece material. A combination of frictional and adiabatic heating leads to softening and allows the tool to penetrate until the shoulder comes into contact with the surface of the workpiece. Welding may be accomplished by traversing the tool along abutting edges of the restrained materials that are to be joined, so that metal flow around the pin leads to coalescence and the formation of a solid-state bond between similar or dissimilar materials. In FSP, the tool may be traversed in a predetermined pattern to process a volume in the workpiece defined by the pin tool profile and the traversing pattern. When applied to a cast metal, FSP may convert the as-cast stir zone (SZ) microstructure to a wrought condition, thereby improving physical and mechanical properties in the absence of a macroscopic shape change. (Swaminathan *et al.*, 2009).

Aluminium alloys are attracting considerable interest worldwide because of their low density, high ratio of strength to weight, high thermal conductivity and good corrosion resistance; however their poor wear resistance causes some limitations for their applications. Metal matrix composites (MMCs) are novel materials with superior mechanical and tribological properties. For use in tribological applications, metal–matrix composites must be able to support a load without undue distortion, deformation, or fracture during performance.

Presence of ceramic particles can provide these needs. Hybrid MMCs are engineering materials that include two or more different reinforcements in order to achieve the combined advantages of them. One of the methods for fabricating in-situ hybrid nano composites is friction stir processing. Nano composites that are produced with FSP have superior properties than those produced with conventional methods such as mechanical alloying, casting, rapid solidification, combustion synthesis, etc. These properties include obtaining a dense solid without porosity, homogeneous distribution of reinforcement particles in matrix and strong bonding between reinforcements and matrix as a result of reaction between them due to the thermo-mechanical condition during FSP. (Anvari *et al.*, 2005)

The strong demand for weight reduction in car and aircraft fabrication urges the optimization of the design of products employing low weight materials. Aluminium and its alloys are used extensively in aerospace and automotive industries because of its low density and high strength to weight Ratio. For many applications, the useful life of components often depends on their surface properties such as wear resistance. Recently, much attention has been paid to friction stir processing (FSP), which is known as a surface modification technique. FSP is a new solid-state processing technique for microstructural modification, which was developed based on the principle of friction stir welding (FSW). The basic concept of FSP is remarkably simple. A rotating tool with pin and shoulder is inserted into a single piece of material and traversed along the desired path to cover the region of interest. Friction between the shoulder and workpiece results in localized heating which raises the temperature of the material to the range where it is plastically deformed. During this process, the severe plastic deformation and thermal exposure of material, results in a significant evolution in the local microstructure. FSP can be accurately described as a forging and extrusion, or metalworking, process. The resulting microstructure is composed of three primary zones: the heat-affected zone (HAZ), the thermo mechanically- affected zone (TMAZ), and the nugget zone (NZ). As a result of intense plastic deformation at elevated temperature, grain refinement to a size range from 0.8 to 12μm can be achieved. It is well known that the nugget zone consists of fine and equiaxed grains produced due to the dynamic recrystallization. FSP has proven to be successful in the modification of various properties such as formability, hardness, yield strength, fatigue and corrosion resistance. (Zahmatkesh *et al.*, 2010)

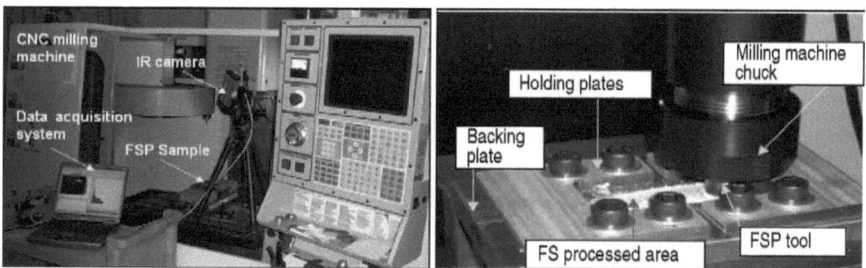

Fig.1.8 Experimental set-up. (Darras *et al.*, 2007)

The difficulty of making high-strength, fatigue and fracture resistant welds in aerospace aluminium alloys, such as highly alloyed 2XXX and 7XXX series, has long inhibited the wide use of welding for joining aerospace structures. These aluminium alloys are generally classified as non-weldable because of the poor solidification microstructure and porosity in the fusion zone. Also, the loss in mechanical properties as compared to the base material is very significant. These factors make the joining of these alloys by conventional welding processes unattractive. Some aluminium alloys can be resistance welded, but the surface preparation is expensive, with surface oxide being a major problem. Friction stir welding (FSW) was invented at The Welding Institute (TWI) of UK in 1991 as a solid-state joining technique, and it was initially applied to aluminium alloys. The basic concept of FSW is remarkably simple. A non-consumable rotating tool with a specially designed pin and shoulder is inserted into the abutting edges of sheets or plates to be joined and traversed along the line of joint. The tool serves two primary functions: (a) heating of workpiece, and (b) movement of material to produce the joint. The heating is accomplished by friction between the tool and the workpiece and plastic deformation of workpiece. The localized heating softens the material around the pin and combination of tool rotation and translation leads to movement of material from the front of the pin to the back of the pin. As a result of this process a joint is produced in 'solid state'. Because of various geometrical features of the tool, the material movement around the pin can be quite complex. During FSW process, the material undergoes intense plastic deformation at elevated temperature, resulting in generation of fine and equiaxed recrystallized grains. The fine microstructure in friction stir welds produces good mechanical properties.

Recently friction stir processing (FSP) was developed as a generic tool for microstructural modification based on the basic principles of FSW. In this case, a rotating tool is inserted in a monolithic workpiece for localized microstructural modification for specific property enhancement. For example, high-strain rate superplasticity was obtained in commercial 7075Al alloy by FSP. Furthermore, FSP technique has been used to produce surface composite on aluminium substrate, homogenization of powder metallurgy aluminium alloy, microstructural modification of metal matrix composites and property enhancement in cast aluminium alloys. FSP is emerging as a very effective solid-state joining/processing technique. In a relatively short duration after invention, quite a few successful applications of FSW have been demonstrated. (Mishra *et al.,* 2005)

Furthermore, the FSP technique has been used for the fabrication of a surface composite on aluminium substrate and the homogenization of powder metallurgy (PM) aluminium alloys, metal matrix composites, and cast aluminium alloys. Compared to other metalworking techniques, FSP has distinct advantages. First, FSP is short-route; solid-state processing technique with one-step processing that achieves micro structural refinement, densification, and homogeneity. Second, the microstructure and mechanical properties of the processed zone can be accurately controlled by optimizing the tool design, FSP parameters, and active cooling/ heating. Third, the depth of the processed zone can be optionally adjusted by changing the length of the tool pin, with the depth being between several hundred micrometers and tens of millimeters; it is difficult to achieve an optionally adjusted processed depth using other metal working techniques. Fourth, FSP is a versatile technique with a comprehensive function for the fabrication, processing, and synthesis of materials. Fifth, the heat input during FSP comes from friction and plastic deformation, which means FSP is a green and energy-efficient technique without deleterious gas, eradiation, and noise. Sixth, FSP does not change the shape and size of the processed components.

Tool geometry is the most influential aspect of process development. The tool geometry plays a critical role in material flow and in turn governs the traverse rate at which FSW can be conducted. An FSP tool consists of a shoulder and a pin. As mentioned earlier, the tool has two primary functions: (a) localized heating, and (b) material flow. In the initial stage of tool plunge, the heating results primarily from the friction between pin and work piece. Some additional heating results from deformation of material. The tool is plunged till the shoulder touches the work piece. The friction between the shoulder and work piece results in the biggest component of heating. From the heating aspect, the relative size of pin and shoulder is important, and the other design features are not critical. The shoulder also provides confinement for the heated volume of material. The second function of the tool is to 'stir' and 'move' the material. The uniformity of microstructure and properties as well as process loads are governed by the tool design. Generally a concave shoulder and threaded cylindrical pins are used. With increasing experience and some for multi pass FSP, conventional cylindrical threaded pin resulted in excessive thinning of the top sheet, leading to significantly reduced bend properties. Furthermore, for lap welds, the width of the weld interface and the angle at which the notch meets the edge of the weld is also important for applications where fatigue is of main concern. Recently, two new pin geometries—Flared-Trifute TM with the flute lands being flared out and A-skew TM with the pin axis being slightly inclined to the axis of machine spindle were developed for improved quality of processing. (Ma *et al.*, 2008)

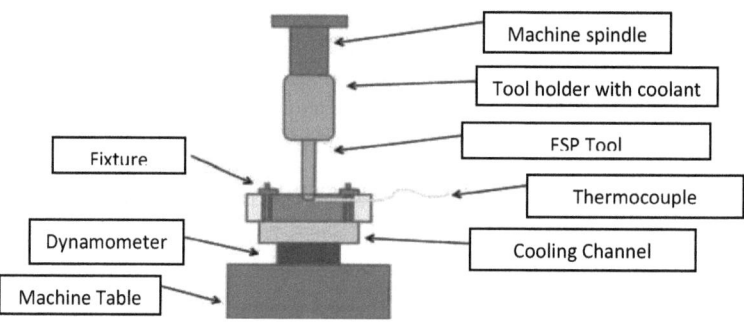

Fig 1.9 shows FSP setup

Recently, friction stir processing technology has been used in aerospace, automotive, marine, and railroad industries along with various other applications. Experimental, analytical and computational studies of friction stir processing nano composites have been performed. Specifically, the effects of processing parameters on microstructure evolution, deformation behaviours and mechanical properties are investigated. Polymers have much lower processing temperatures than metals. The facilities for implementing FSP could be much simpler and less expensive. The polymeric materials used are particle filled plastics such as polystyrene (PS) and nylon.

Friction stir processing (FSP) forces materials flow at the temperature lower than the melting temperature. Materials are extruded, forged, consolidated and cooled under hydrostatic pressure conditions. Thus some of the properties, for example, hardness is high for FSP materials. The primary research on friction stir processing focuses on aluminium alloys.

Applying this technology for processing other alloys and materials including stainless steels, magnesium, titanium, and copper are also studied. This technology has found applications in modifying the microstructure of reinforced metal matrix composite materials. It is also used in processing polymeric composite materials. It is possible to transfer FSP technology into energy materials field for example, for fusion reactor materials including vanadium alloy joining and processing. There are some major challenges for improving the FSP technology. One is the tool wear in processing reinforced composite materials. The other challenge is how to increase the joining strength and improve the fatigue property of FSP composite materials. (Gan *et al.*, 2010)

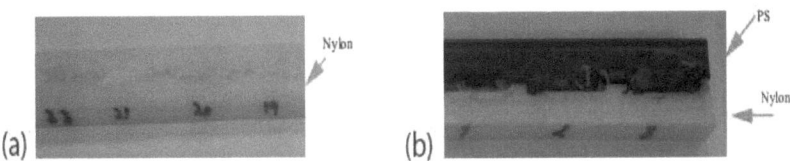

Fig 1.10 (a) FSP nylon-nylon laminate and (b) shows nylon-polystyrene laminate+

1.2 FSP Parameters

FSP/FSW involves complex material movement and plastic deformation. Welding parameters, tool geometry, and joint design exert significant effect on the material flow pattern and temperature distribution, thereby influencing the microstructural evolution of material. A few major factors affecting FSP process, such as tool geometry, welding parameters, joint design are addressed.

1.2.1 Tool geometry

Tool geometry is the most influential aspect of process development. The tool geometry plays a critical role in material flow and in turn governs the traverse rate at which FSW can be conducted. An FSP tool consists of a shoulder and a pin as shown schematically in Fig 1.11. As mentioned earlier, the tool has two primary functions: (a) localized heating, and (b) material flow. In the initial stage of tool plunge, the heating results primarily from the friction between pin and workpiece. Some additional heating results from deformation of material. The tool is plunged till the shoulder touches the workpiece. The friction between the shoulder and workpiece results in the biggest component of heating. From the heating aspect, the relative size of pin and shoulder is important, and the other design features are not critical.

The shoulder also provides confinement for the heated volume of material. The second function of the tool is to 'stir' and 'move' the material. The uniformity of microstructure and properties as well as process loads are governed by the tool design. Generally a concave shoulder and threaded cylindrical pins are used.

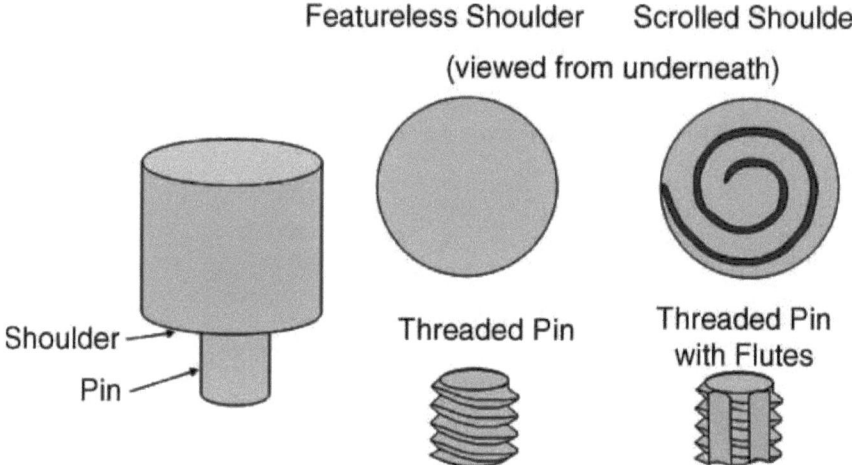

Fig 1.11 Schematic drawing of the FSP tool (Mishra *et al.*, 2005)

1.2.2 Processing parameters

For FSP, two parameters are very important: tool rotation rate (v, rpm) in clockwise or counterclockwise direction and tool traverse speed (n, mm/min) along the line of joint. The rotation of tool results in stirring and mixing of material around the rotating pin and the translation of tool moves the stirred material from the front to the back of the pin and finishes welding process. Higher tool rotation rates generate higher temperature because of higher friction heating and result in more intense stirring and mixing of material as will be discussed later. However, it should be noted that frictional coupling of tool surface with workpiece is going to govern the heating. So, a monotonic increase in heating with increasing tool rotation rate is not expected as the coefficient of friction at interface will change with increasing tool rotation rate.

1.2.3 Tool Tilt angle

In addition to the tool rotation rate and traverse speed, another important process parameter is the angle of spindle or tool tilt with respect to the workpiece surface. A suitable tilt of the spindle towards trailing direction ensures that the shoulder of the tool holds the stirred material by threaded pin and move material efficiently from the front to the back of the pin. Further, the insertion depth of pin into the workpieces (also called target depth) is important for producing sound appearance with smooth tool shoulders. The insertion depth of pin is associated with the pin height. When the insertion depth is too shallow, the shoulder of tool does not contact the original workpiece surface. Thus, rotating shoulder cannot move the stirred material efficiently from the front to the back of the pin, resulting in generation of welds with inner channel or surface groove. When the insertion depth is too deep, the shoulder of tool plunges into the workpiece creating excessive flash. In this case, a significantly concave appearance is produced, leading to local thinning of the processed plate. It should be noted that the recent development of 'scrolled' tool shoulder allows FSP with 0' tool tilt.

1.3 Process modelling

FSP results in intense plastic deformation and temperature increase within and around the stirred zone. This results in significant microstructural evolution, including grain size, grain boundary character, dissolution and coarsening of precipitates, breakup and redistribution of dispersoids, and texture. An understanding of mechanical and thermal processes during FSP is needed for optimizing process parameters and controlling microstructure and properties of welds.

1.3.1. Metal flow

The material flow during friction stir welding is quite complex depending on the tool geometry, process parameters, and material to be welded. It is of practical importance to understand the material flow characteristics for optimal tool design and obtain high structural efficiency welds. This has led to numerous investigations on material flow behavior during FSP. A number of approaches, such as tracer technique by marker, welding of dissimilar alloys/metals, have been used to visualize material flow pattern in FSP. In addition, some computational methods including FEA have been also used to model the material flow.

1.3.2. Temperature distribution

FSP results in intense plastic deformation around rotating tool and friction between tool and workpieces. Both these factors contribute to the temperature increase within and around the stirred zone. Since the temperature distribution within and around the stirred zone directly influences the microstructure of the welds, such as grain size, grain boundary character, coarsening and dissolution of precipitates, and resultant mechanical properties of the welds, it is important to obtain information about temperature distribution during FSP. However, temperature measurements within the stirred zone are very difficult due to the intense plastic deformation produced by the rotation and translation of tool. Therefore, the maximum temperatures within the stirred zone during FSP have been either estimated from the microstructure of the processed zone or recorded by embedding thermocouple in the regions adjacent to the rotating pin.

1.4 Microstructural evolution

The contribution of intense plastic deformation and high-temperature exposure within the stirred zone during FSP results in recrystallization and development of texture within the stirred zone and precipitate dissolution and coarsening within and around the stirred zone. Based on microstructural characterization of grains and precipitates, three distinct zones, stirred (nugget) zone, thermo-mechanically affected zone (TMAZ), and heat-affected zone (HAZ), have been identified as shown in Fig. 1.12. The microstructural changes in various zone have significant effect on post process mechanical properties. Therefore, the microstructural evolution during FSP has been studied by a number of investigators.

Fig 1.12 A typical macrograph showing various microstructural zones in FSP (Mishra *et al.*, 2005)

1.4.1 Nugget zone

Intense plastic deformation and frictional heating during FSP result in generation of a recrystallized fine-grained microstructure within stirred zone. This region is usually referred to as nugget zone (or weld nugget) or dynamically recrystallized zone (DXZ). Under some FSP conditions, onion ring structure was observed in the nugget zone (Fig 1.12). In the interior of the recrystallized grains, usually there is low dislocation density. However, some investigators reported that the small recrystallized grains of the nugget zone contain high density of sub-boundaries, subgrains, and dislocations. The interface between the recrystallized nugget zone and the parent metal is relatively diffuse on the retreating side of the tool, but quite sharp on the advancing side of the tool.

1.4.1.1. Shape of nugget zone

Depending on processing parameter, tool geometry, temperature of workpiece, and thermal conductivity of the material, various shapes of nugget zone have been observed. Basically, nugget zone can be classified into two types, basin-shaped nugget that widens near the upper surface and elliptical nugget.

1.4.1.2 Grain size

It is well accepted that the dynamic recrystallization during FSP results in generation of fine and equiaxed grains in the nugget zone. FSP parameters, tool geometry, composition of workpiece, temperature of the workpiece, vertical pressure, and active cooling exert significant influence on the size of the recrystallized grains in the FSP materials

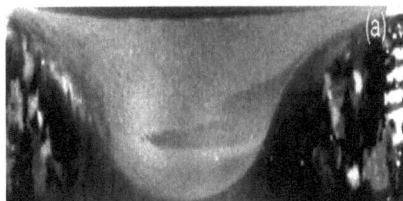 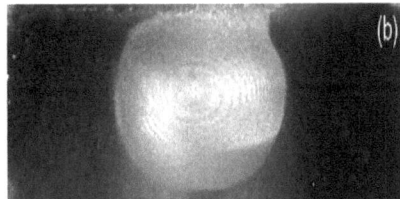

Fig 1.13 Effect of processing parameter on nugget shape in FSP A356: (a) 300 rpm, 51 mm/min and (b) 900 rpm, 203 mm/ min (Mishra *et al.,* 2005)

The tool geometry was not identified in a number of studies. While the typical recrystallized grain size in the FSP aluminium alloys is in the micron range (Table 1.1), ultrafine grained (UFG) microstructures (average grain size <1 mm) have been achieved by using external cooling or special tool geometries.

FSP results in the temperature increase up to 400–550 8C within the nugget zone due to friction between tool and workpieces and plastic deformation around rotating pin. At such a high temperature precipitates in aluminium alloys can coarsen or dissolve into aluminium matrix depending on alloy type and maximum temperature.

Table 1.1 A summary of grain size in nugget zone of FSP aluminium alloys

Material	Plate thickness mm	Tool geometry	Rotation rate (rpm)	Transverse speed (mm/min)	Grain size (μm)
6061Al-T6	6.3	Cylindrical	300–1000	90–150	10
1100Al	6.0	Cylindrical	400	60	4
2024Al	6.35	Threaded, cylindrical	200–300	25.4	2.0–3.9
5083Al	6.35	Threaded, cylindrical	400	25.4	6.0
7075Al-T65	6.35	Threaded, cylindrical	350, 400	102, 152	3.8, 7.5

1.4.2 Thermo-mechanically affected zone

Unique to the FSP process is the creation of a transition zone—thermo-mechanically affected zone (TMAZ) between the parent material and the nugget zone, as shown in Fig.1.14. The TMAZ experiences both temperature and deformation during FSP. The TMAZ is characterized by a highly deformed structure. The parent metal elongated grains were deformed in an upward flowing pattern around the nugget zone. Although the TMAZ underwent plastic deformation, recrystallization did not occur in this zone due to insufficient deformation strain. However, dissolution of some precipitates was observed in the TMAZ, due to high-temperature exposure during FSP. The extent of dissolution, of course, depends on the thermal cycle experienced by TMAZ. Furthermore, it was revealed that the grains in the TMAZ usually contain a high density of sub-boundaries.

Fig 1.14 Microstructure of thermo-mechanically affected zone in FSP (Mishra *et al.*, 2005)

1.4.3 Heat-affected zone

Beyond the TMAZ there is a heat-affected zone (HAZ). This zone experiences a thermal cycle, but does not undergo any plastic deformation. Mahoney et al defined the HAZ as a zone experiencing a temperature rise above 250'C for a heat-treatable aluminium alloy. The HAZ retains the same grain structure as the parent material. However, the thermal exposure above 250'C exerts a significant effect on the precipitate structure.

1.5 Application of FSP

Friction stir welding has a number of attributes that can be used to develop a generic tool for microstructural modification and manufacturing. This has led to several applications for microstructural modification in metallic materials, including superplasticity, surface composite, homogenization of nanophase aluminium alloys and metal matrix composites, and microstructural refinement of cast aluminium alloys.

1.5.1 Superplasticity

It is well known that two basic requirements are necessary for achieving structural superplasticity. The first is a fine grain size, typically less than 15 mm. The second is thermal stability of the fine microstructure at high temperatures. Conventionally, thermo-mechanical processing (TMP) is used to produce fine-grained microstructure in commercial aluminium alloys. A typical TMP for heat-treatable aluminium alloys consists of solution treatment, overaging, multiple pass warm rolling (200–220 8C) with intermittent re-heating, and a recrystallization treatment. Clearly, TMP is complex and time-consuming and results in increased material cost. More importantly, the optimum superplastic strain rate of 1×10^{-4} to 10×10^{-3} s^{-1} obtained in TMP commercial aluminium alloys such as 7075 and 7475 is too slow for superplastic forging/forming of components in the automotive industry. To advance superplastic forming (SPF) into mass production oriented industries, there is a need to develop new processing techniques and/or aluminium alloys to shift the optimum superplastic strain rate to high-strain rate

1.5.2 Surface composites

Compared to unreinforced metals, metal matrix composites reinforced with ceramic phases exhibit high strength, high elastic modulus, improved resistance to wear, creep and fatigue, which make them promising structural materials for aerospace and automobile industries. However, these composites also suffer from a great loss in ductility and toughness due to incorporation of nondeformable ceramic reinforcements, which limits their applications to a certain extent. For many applications, the useful life of components often depends on their surface properties such as wear resistance. In these situations, it is desirable that only the surface layer of components is reinforced by ceramic phases while the bulk of components retain the original composition and structure with higher toughness.

In recent years, several surface modification techniques, such as high-energy laser melt treatment, high-energy electron beam irradiation, plasma spraying, cast sinter, and casting, have been developed to fabricate surface metal matrix composites. Among these techniques, laser melt treatment (also called laser processing or laser surface engineering (LSE)) is widely used for surface modification. However, it should be pointed out that the existing processing techniques for forming surface composites are generally based on liquid phase processing at high temperatures. In this case, it is hard to avoid the interfacial reaction between reinforcement and metal matrix and formation of some detrimental phases. Furthermore, critical control of processing parameters is necessary to obtain ideal solidified microstructure in surface layer. Obviously, if processing of surface composite is carried out at temperatures below melting point of substrate, the problems mentioned above can be avoided. Recently, studies were conducted by Mishra et al. to incorporate ceramic particles into surface layer of aluminium alloy (5083Al and A356) to form surface composite by means of FSP. They reported that the processing parameters (tool geometry, tool rotation rate, traverse speed, and target depth) exhibit significant effects on formation of surface composite layer

1.5.3 Microstructural modification

Various modification and heat-treatment techniques have been developed to refine the microstructure of cast Al–Si–Mg alloys. The first category of research is aimed at modifying the morphology of Si particles. For example, eutectic modifiers such as sodium, strontium, and antimony are widely used to spheroidize Si particles. However, there are some drawbacks with these modifiers. For sodium, the benefits fade rapidly on holding at high temperature and the modifying action practically disappears after only two remelts. For strontium, the density of microshrinkage porosity is increased after the addition of strontium due to owing to increased gas pickup from the dissolution difficulty and a depression in the eutectic transformation temperature. For antimony, environmental and safety concerns have precluded its use in most countries. Alternatively, heat treatment of cast alloys at high temperature, usually at the solid solution temperature around 540 8C for long time, is also used to modify the morphologies of Si particles. Solution heat-treatment results in a substantial degree of spheroidization of Si particles and also coarsens Si particles. However, solution treatment at high temperature for long time increases material cost. The second research category refines the coarse primary aluminium phases. Heat treatment at an extremely high temperature of 577 8C for a short time of 8 min resulted in a substantial refinement in the aluminium dendrites in a semi-solid processed (SSP) A356. Furthermore, it was reported that a melt thermal treatment led to a remarkable refinement of the aluminium phase in A356, thereby resulting in a significant improvement in both strength and ductility. It is important to point out that none of the modification and heat-treatment techniques mentioned above can eliminate the porosity effectively in Al–Si–Mg castings and redistribute the Si particles uniformly into the aluminium matrix. As presented above, during FSP, tool transports materials from the front to the back of the tool in a complex way, resulting in intense deformation and mixing of material. It is expected that such a process can refine effectively the microstructure of Al–Si–Mg castings. Recently, Ma et al. investigated the effect of FSP on microstructure and properties of A356. FSP resulted in a significant breakup of coarse acicular Si particles and primary aluminium dendrites, created a homogeneous distribution of Si particles in the aluminium matrix, and nearly eliminated all casting porosity. These microstructural modifications significantly improved the mechanical properties of cast A356, in particular ductility and fatigue lifetime. (Mishra *et al.*, 2005)

1.6 Need for present study

Aluminium is the most popular metal that is widely used. About 85% of aluminium is used for wrought products, for example rolled plate, foils and extrusions. Aluminium has light weight, resistance to corrosion and has low melting point but its severe limitation is the difficulty associated with welding of aluminium/aluminium alloy structures. Further, the low hardness and strength of aluminium or its alloys also limit their use especially for tribological applications. The research work carried out by some of the investigators reveals that friction stir processing technique can be successfully used for microstructural modification, enhancement of surface properties such as micro hardness, wear resistance etc. of aluminium or its alloys. Most of the research work pertaining to aluminium/aluminium alloy is dedicated to investigate the microstructural changes and enhancement of mechanical properties achieved by friction stir welding (FSW). The work pertaining to enhancement of surface properties aluminium/aluminium alloys through friction stir processing (FSP) is rather scant.

In this light it is divided to investigate the microstructure and mechanical properties e.g. microhardness and impact strength, of 6063 series of aluminium alloy. Still there is a lot of scope to carry out the the work related to enhance the surface properties of a material through friction stir processing (FSP).

1.7 Chapter Scheme

Chapter 1:   Introduction

In this chapter the introduction about the whole topic FSP has been given. The introduction consists of all the important parameters regarding the FSP process. The chapter consists about the tool used for the process, different tool design, and general setup

Chapter 2:   Literature Review

In this chapter the different research work carried out by different researchers and their results are discussed in brief. In this the results obtained after experiments on different material and their parameters are discussed.

Chapter 3:   Experimental Procedure

In this chapter the experimental detail about the present study is described. In this chapter detail is given about the conducted experiment with the selected parameters and about the testing after experiment is discussed.

Chapter 4:   Results and Discussion

In this chapter the detail about the result obtained after testing of the experimented samples and their effects are discussed. It contains about microstructure evaluation, hardness and impact test conducted on the samples.

Chapter 5:   Conclusions and Future scope

In the last chapter the results obtained after the testing are compared with each other and conclusion is made after discussing and checking all parameters and discussed about the advancement of the study in the future.

CHAPTER 2
LITERATURE REVIEW
2.1 Review of related literature
Following is the literature review of Friction Stir Processing Technique:

Akramifard *et al.,(2013)* studied the effect of fabrication via friction stir processing on microstructure and mechanical properties of Cu/SiC metal matrix composite. Commercially pure Cu sheets with the dimensions of 100 mm x 70 mm x 5 mm were used as base metal. SiC particles (25 µm) were used as reinforcement. The constant rotational and travel speeds were 1000 rpm and 50 mm/min. SZ has fine and equiaxed grains and distribution of SiC particles in the matrix is almost uniform. Stired zone showed increased microhardness and better wear behaviour.

Anvari *et al.,(2013)* studied the Wear characteristics of Al–Cr–O surface nano-composite layer fabricated on Al6061 plate by friction stir processing. Commercially available Al6061-T6 plates of 13 mm thickness were used as a base material. The optimum processing parameters for FSP included a traveling speed of 100 mm/min and a rotational rate of 630rpm with a tilt angle of 3' A new procedure was introduced for applying the reinforcement particles, in which Cr2O3 powder was applied on Al6061 plate by atmosphere plasma spray and then the FSP for six passes was performed on the plate. A homogenous distribution of reinforcement particles over the nugget zone was produced by FSP without any defects. FSP reduced the wear resistance of Al6061-T6 without reinforcement due to the loss of hardening precipitates during process. The wear mechanism for nano-composite was different from as-received Al and FSPed samples. Adhesive and abrasive are the dominant wear mechanisms for as-received Al6061-T6 and FSPed samples, while results showed delamination wear for the nano-composite. Dispersion of reinforcement particles as hard ceramic phase in Al improves the wear resistance

Aruri *et al.,(2011)* studied the effect of 3 Pass FSP on of AA6061 /SiCp Surface Composite Fabricated by Friction Stir Processing. Commercial SiC particles (average size: 20µm) and AA6061 rolled plate (thickness: 4mm) were used to fabricate composite by FSP. A groove was prepared with dimensions of 3mm width and 3mm depth on the Al alloy. The FSP tool was made of H13 tool steel and had a cylindrical shape shoulder (24mm) with a screwed taper pin profile (8 mm) In the beginning of the FSP, the groove was filled with SiC particles and covered with a modified FSP tool that only had a shoulder without pin to prevent the SiC particles from being displaced out of the groove. The FSP parameters such as tool rotational speed and travelling speed were 1400rpm, 40mm/min respectively selected. It is observed that after the first and third pass of FSP the SiC particles were uniformly distributed in the stir zone without any defect than. This was due to the stirring action generated in each pass by the rotated tool. There is no reaction between SiCp and the base metal because of solid-state process. In third pass the reinforced particles were easily wrapped by softening the metal and rotated with the FSP tool rather than the first pass. It observed that microhardness of the first pass of FSPed composite in nugget zone is higher than that of the as-received Al alloy which was measured to be 104Hv. It is considered that higher value was obtained due to the pinning effect and presence of hard SiC particles. Where as in third pass of FSPed composite, the microhardness was lower compared to first pass and as-received Al alloy this is due to the material become more softening and annealing effect of heat input during the process. In static immersion corrosion test the FSP AA6061 /SiCp exhibited significantly greater corrosion resistance in I-pass than compared to the III-pass and as-received Al alloy.

Bauri *et al.,(2011)* studied the effect of friction stir processing (FSP) on microstructure and properties of Al–TiC in situ composite. The as-cast composite plate was machined to a thickness of 10mm for friction stir processing (FSP). The plate was subjected to a single and a double-pass FSP on two separate tracks. The tool is made of M2 tool steel and essentially consists of a shoulder and a pin. The shoulder diameter is 12mm and the pin is 4mm in diameter and 3.5mm in length. A rotational speed of 1000rpm and traverse speed of 60mm/min was used. Microstructure characterization, Hardness and tensile tests were carried out to assess the effect of FSP on mechanical properties of the Al–TiC composite. Hardness of base material was 38 Hv and after single pass it increased to 48 Hv and after double pass it further increased to 58 Hv. The improved homogeneity of particle distribution after FSP gives rise to more effective dispersion hardening. The grain refinement and high dislocation density also contribute to enhancement in the hardness. The tensile strength of base material was 88mpa, it increased to 103 mpa after 1pass, and further increased to 123mpa after 2 pass. Better homogeneity of particle distribution gives rise to better dispersion strengthening. A finer grain size is achieved after double-pass FSP and this gives rise to further enhancement in strength.

Cavalierea *et al.*,(2005) investigated the tensile mechanical properties was evaluated of 7075 AL alloy at room temperature and higher temperature and different strain rates in nugget zone to analyze the superplastic properties of the recrystallized material to observe the differences from the parental material as the function of strong grain refinement due to friction stir processing, revealed the classical formation of elliptical onion structure at the Centre of the specimen, this is a structure characterized by fine and equiaxed recrystallized. The micro-hardness reaches a value of 124 Hv in the centre and after increasing in both the sides it starts to decrease after a maximum (141 Hv) at a distance of 3.5 mm from the centre the mechanical results are very good considering the drastic conditions to which the material is subjected during the Friction Stir Process, the elastic modulus results are very different with respect to the parent material, The strength and ductility of the material increased strongly in the nugget zone with respect to the material in the transverse direction. The very high strain to fracture values reached by the studied material at high temperatures result in the very attractive superplastic behaviour exhibited by such alloys, Microscopic voids observed in the specimen tested at 300'C. The fracture surfaces of the specimens tested at different conditions of temperatures and strain rates were extensively investigated by using a FEGSEM microscope revealing the defects resulting from the Friction process and the microscopic mechanisms taking place during hot deformation. The results have shown the retaining high mechanical properties with respect to parent metal.

Cavaliere *et al.*,(2006) studied the effect of Friction stir process on AZ91 magnesium alloy produced by high pressure die cast. The material was produced by HPDC into the form of special trials of 2.5mm thickness. The material was solution treated at 415◦C for 2 h in order to soften the alloy before tool processing. The material was subjected to friction stir processing by employing a flat C60 steel tool with rotating speed of 700 rpm and a travelling speed of 2.5 mm/s. The material was friction stir processed showing good strength and ductility values at room temperature because of the very fine structure obtained by the processing. The FSP produces a strong increase in mechanical properties respect to the unstirred material with an increase in the elongation to failure accompanied with a strong increase in strength due to the very fine recrystallized structure and to the absence of porosity produced by the stirring process. In the HPDC as received material, in fact, large porosities and oxides formation were observed on the fracture surfaces. Such voids disappear after FSP and no oxides presence was observed on the fracture surfaces. The oxides particles are, in fact, broken in a very fine shape from the tool action and reduced to a very smaller dimension respect to than in the case of as-cast material.

Choi et al.,(2013) investigated the microstructure and mechanical properties of A356 based composites using friction stir process. The material used in this study was a piece of A356 Al alloy with dimensions of 140 mm×70 mm×4 mm. Microstructural observations were carried out at the cross sections perpendicular to the FSP direction by optical microscopy (OM) and scanning electron microscopy (SEM). The Vickers hardness profile of the stir zone (SZ) was measured. The mechanical properties of the SZ with SiC particles, compared to the BM and SZ without SiC, were improved by the dispersed Si, SiC particles and the homogeneous microstructure.

Darras et al.,(2010) conducted experiment on commercial AZ31 magnesium alloy. He conducted experiment on 3 samples, 1 as unprocessed sample, 2 sample is processed at rpm 1200 and feed 22inch./min., and on 3rd sample rpm was same but feed rate was increased to 25inch/min, he observed FSP produced more homogeneous microstructure. The unprocessed sample has grain size of 6μm, the 2nd sample showed grain refinement of grain size 3-4μm. He observed that increase in rotational speed decreases the hardness.

Deepak et al.,(2013) prepared the 5083 Al-SiC surface composite by FSP and checked its mechanical properties. Sample of 5083-Al alloy in plate form (Length: 150 mm, Width: 100 mm, Thickness: 6 mm) has been selected for the purpose of carrying out FSP. Microstructure, wear test and microhardness is checked out. The micro hardness is increased from 44HV to 84 HV. The wear resistance of FSPed sample is inferior to that observed for 5083Al in spite of its higher hardness.

Devaraju et al.,(2013) investigated the effect of rotational speed and reinforcement particles such as silicon carbide (SiC), alumina (Al2O3) on wear and mechanical properties of aluminium alloy based surface hybrid composites fabricated via friction stir processing. The base material employed in this study is 4 mm thick aluminium alloy 6061. The tool traveling speed of 40 mm/min, axial force of 5 KN and tool onward tilt angle of 2.5' along the centre line were used in FSP. Microstructure was checked of stirred zone, mechanical properties such as wear, tensile and microhardness was checked. Microhardness was increased, wear and tensile properties decreased as compared to the base metal.

Elangovan *et al.,(2008)* studied the influences of tool pin profile and axial force on the formation of friction stir processing zone in AA6061 aluminium alloy. The rolled plates of 6 mm thickness, AA6061 aluminium alloy were cut into the required sizes (300×150 mm) by power hacksaw cutting and grinding. Square butt joint configuration was prepared to fabricate FSW joints. Non-consumable tools made of high carbon steel were used to fabricate the joints. An indigenously designed and developed machine (15 HP; 3000 RPM; 25 kN) was used to fabricate the joints. Five different tool pin profiles were used to fabricate the joints. Using each tool, three joints at three different axial force levels and in total 15 joints (5×3) were fabricated in this investigation. All the joints fabricated in this investigation were analysed at low magnification (10X) using optical microscope to reveal the quality of FSP regions. Out of the three joints fabricated using straight cylindrical pin profiled tool only the joint fabricated at an axial force of 6 kN is found to be defective. Similarly, in the case of joints fabricated by tapered cylindrical pin profiled tool the joint fabricated at an axial force of 6 kN is found to be defective. In contrast, the joints fabricated by threaded cylindrical pin profiled tool square pin profiled tool and triangular pin profiled tool are found to be free from any kind of defects, irrespective of axial force applied. From the macrostructure analysis, it can be inferred that the formation of defect free FSP zone is a function of tool profile and axial force applied. Of the five tool pin profiles used in this investigation to fabricate the joints, square pin profiled tools produce defect-free, good quality FSP region, irrespective of applied axial force levels. Of the 15 joints fabricated in this investigation, the joint fabricated using the square pin profiled tool at an axial force of 7 kN showed superior tensile properties. A defect-free FSP region, smaller grains with uniformly distributed finer strengthening precipitates in FSP region and higher hardness are the reasons for superior tensile properties of the above joints.

Giles *et al.,(2008)* studied the Effect of Friction Stir Processing on the microstructure and mechanical Properties of an Aluminium Lithium Alloy. For this study, two plates, each approximately 580 X 380 X 13 mm in size, were sectioned from rolled AA2099 plate. The tool had a shoulder 25 mm in diameter and a threaded pin 12.7 mm in diameter and 12.5 mm in depth. Processing was conducted at 400 revolutions per minute (rpm) and a traversing speed of 127 mm m-1. The elongated pancake like grains typically produced in conventional thermo mechanical processing of this alloy are replaced by equiaxed grains during FSP. After FSP, the hardness decreases steeply from a value of 57 HRB in the stir zone near the surface in contact with the tool shoulder to a value of 23 HRB close to the bottom of the stir zone.

Upon annealing for 24 hours to re age the material after FSP, the hardness of the stir zone decreases further to a value of ~30 HRB at the top surface while maintaining a hardness value of ~23 HRB at the bottom. On further increasing the annealing time, the hardness increases uniformly throughout the plate. Thus, after 72 hours annealing following FSP, the hardness value close to the surface is ~73 HRB while the hardness close to the bottom of the stir zone is ~59 HRB. No appreciable difference in yield or tensile strength was seen.

Hsu *et al.*,(2005) prepared Ultrafine-grained Al–Al2Cu composite produced in situ by friction stir processing. The starting materials used are pure aluminium powder (99.7% purity, _325 mesh) and pure copper powder (99.5% purity, _320 mesh). The copper content of the alloy is 15 at. % (denoted as Al–15Cu). The premixed Al–15Cu alloy powders were cold compacted to a small billet (12 x 12 x 88 mm) in a steel die, set by using a pressure of 225 MPa. To improve the billet strength for easier handling in FSP, the green compact was sintered for 20 min in air at either 773 K or 803 K. The tool pin used is standard M1.2*6. A counter clockwise tool rotation speed of 700 rpm was used, and the rotating tool was traversed at a speed of 45 mm/min along the long axis of the billet. In order to obtain a fully dense solid from a powder compact, two FSP passes were applied to the billet. The FSP resulted in significant increase in hardness from 80 Hv in BM to 160 ± 14 Hv in the SZ. The microstructure observations indicated that Cu nearly completely reacted with Al to form fine Al2Cu particles in a short FSP time. This work has demonstrated that Al2Cu intermetallic reinforced aluminium matrix composites with ultrafine grained structure can be fabricated in situ by FSP. The Al–Al2Cu composites thus formed are fully dense, and the Al2Cu particles are distributed quite homogeneously

Hsu *et al.*,(2006) investigated the properties of Al–Al3Ti nano composites produced in situ by friction stir processing. The starting materials used were aluminium powder and titanium powder. Titanium contents of 5, 10, or 15 at. % were pre-mixed with aluminium powder. The pre-mixed Al–Ti alloy powders were cold compacted to 12 X 20 X 88 mm billets in a steel die set using a pressure of 225 MPa. To improve the billet strength for easier handling in FSP, the green compact was sintered for 20 min in air at 823 K. The tool pin used in FSP is a standard M1.2*6. Counter clockwise tool rotation with a speed of 700 or 1400 rpm was used, and the rotating tool was traversed at a speed of 45 mm/min along the long axis of the billet. XRD was used to identify the phases present in the SZ of specimens during FSP. The diffraction patterns showed that Ti reacted with Al to form Al3Ti, but some unreacted Ti remained after four FSP passes. After FSP, the average size of Ti particles is refined from 40 μm to about 1–5 μm. For Al–15Ti after 4 FSP passes, the volume fraction of Al3Ti is close to 0.5, which results in a hardness value of 200 Hv. Typical microstructures which shows a large number of fine Al3Ti particles uniformly dispersed in an ultrafine-grained Al matrix. The fine Al3Ti particles were found within the grain interior as well as along the grain boundaries of the Al matrix. The Al3Ti particles were identified by the use of electron diffraction.

Jiang *et al.*,(2011) studied the effect of NANO-SiO2 particles reinforced magnesium alloy produced by friction stir processing. Friction stir processing (FSP) was employed to produce SiO2/AZ31 composites for increase of the hardness of AZ31 matrix. The metal matrix used wasAZ31magnesium alloy with Nano-SiO2 particles with average sizes of ~20 nm were commercially available and used as reinforcement particles. AZ31 Mg alloy was machined to a plate and a groove with the dimensions of width 3mmand depth 2.5 mm in center of a surface tested. First nano-SiO2 particles were filled in the groove. Then the probe was inserted into the groove filled by the nano-SiO2 particles and moved with a rotating rate of 1200 rpm and a travel speed of 50 mm/min. The microstructures were observed by optical microscopy (OM), and scanning electron microscopy (SEM). SEM was employed to examine the microstructures with particles dispersion. Vickers hardness tests were conducted using a 300 g load for 15 s on the surface of the plate before and after FSP. Nano-SiO2 particles were successfully dispersed into AZ31 magnesium alloy via friction stir processing (FSP). The microstructures and microhardness changes before and after FSP were investigated by OM, SEM and Vickers hardness tests.SiO2 particles were uniformly dispersed intoAZ31 matrix after FSP with sizes of less than 0.2 μm. New grains size evolved in composite zone is less than 1 um and much finer than that in the regions outside the stirred zone. Hardness of the SiO2/AZ31 composites was 90 Hv and is about 1.83 times higher than that of the as-receivedAZ31.

Kurt *et al.*,(2011) investigated the surface modification of aluminium by friction stir processing. The starting materials were cold-rolled plates of 1050 aluminium alloy. The surface of plates was cleaned with grinding paper before processing. The average size of the SiC particles was about 10_m. SiC particles were added into a small amount of methanol and mixed, and then applied to surface of the plates to form a thin SiC particle layer. No binder was added to the mixture. The dimensions of the workpiece were 50mm×100mm×5mm. The tool was made from AISI 1050 steel. The tool was rotated clockwise at rotation speed of 500–700–1000 rpm, with the rotating shoulder 0.1mm inserted into the workpiece. The travelling speeds were 15–20–30 mm/min. Microstructural observations and Vickers microhardness test were performed on the treated surfaces. FSP decreased the grain size and increased the hardness of processed material. Increased rotation speed and low travelling speeds caused more heat input which affects the thickness of the surface layer, grain size and distribution of the precipitates and reinforcing particles. A good dispersion of SiCp can be obtained for the composite layer produced by parameters 1000rpm and 20 mm/min.

Lee *et al.*,(2008) tested mechanical properties of Al–Fe in situ nanocomposite produced by friction stir processing. Al–10Fe powders were cold compacted to a small billet (12*12*88 mm) in a steel die set by using a pressure of 225 MPa. To improve the billet strength for easier handling in FSP, the green compact was sintered in Air atmosphere at 823 K for 20 min. the reaction was not processed after sintering but the reaction started after FSP the hardness of the processed metal increases but the reaction was not complete even after the four passes of FSP. The reinforcing particles were identified as Al13Fe4.

Mahmouda *et al.*,(2010) investigated the wear characteristics of surface-hybrid MMCs layer fabricated on aluminium plate by friction stir processing. Commercially pure aluminium Al-1050-H24 plates of 5mm thickness were used as the base material. Mixtures of SiC and Al2O3 particles at different ratios were used as the reinforcements. The reinforcement powder was packed in a groove of 3mm width and 1.5mm depth cut on the Al plate. An aluminium sheet of 2mm thickness was used to cover the groove filled with the reinforcement powders to prevent them from going out before they were incorporated into the aluminium matrix during the FSP. The tool was rotated at a rotation speed of 1500 rpm, and travelled at a speed of 1.66mm/s with a tilt angle of 3°. Macroscopic appearances of the nuggets cross-sections produced by triple FSP passes with different relative contents of SiC and Al2O3 powders. The reinforcement particles (SiC, Al2O3 or their mixture) were distributed almost homogenously over the nugget zone by FSP without any defects except small some voids forming around the Al2O3 particles. The average hardness of the resulted composites increased to about 60HV at 100% SiC and it decreased with increasing the relative ratio of Al2O3 particles. At a normal load of 2N, the wear resistance decreased with increasing the Al2O3 particles ratio in the reinforcement. However, at a normal load of 5N, the hybrid composites containing 20% Al2O3 + 80% SiC exhibited superior wear resistance to other relative ratios of the Al2O3 and SiC particles. The wear resistance at normal load of 10N was almost independent of the relative ratios of the Al2O3 and SiC particles, and it was close to the monolithic FSP sample.

Morisada *et al.,(2006)* in this study commercially available MWCNTs (outer diameter: 20–50 nm, length:~250 nm) synthesized from hydrocarbons, and an AZ31 rolled plate (thickness: 6 mm) were used. The MWCNTs were filled into a groove (1mm×2 mm) on the AZ31 plate before the FSP was used. The FSP tool made of SKD61 has a columnar shape (Ø 12 mm) with a probe (Ø 4 mm, length: 1.8 mm). A constant tool rotating rate of 1500 rpm was adopted and the travel speed was varied from 25 to 100 mm/min. OM and SEM images obtained from the surface composites fabricated by the FSP, respectively. The dispersion of the MWCNTs in the AZ31 matrix was related to the travel speed of the rotating tool. Entangled MWCNTs, which were similar to the as-received MWCNTs, could be observed in the sample FSPed at 100 mm/min. Though the sample FSPed at 50 mm/min showed a better dispersion of the MWCNTs, there were some regions which included the aggregated MWCNTs. On the other hand, a good dispersion of the MWCNTs, which were separated from each other, could be observed for the sample FSPed at 25 mm/min. Only the travel speed of the rotating tool determined the friction heats in the matrix because the tool rotating rate was constant (1500 rpm) in this study. It is considered that the travel speed of 100 mm/min was too fast to produce enough heat flow to produce a suitable viscosity in the AZ31 matrix for the dispersion of the MWCNTs. The FSP with MWCNTs obviously increases the microhardness of the substrates. The maximum microhardness for the composites is 78 Hv, while that of the sample treated by the FSP without MWCNTs and the as-received sample are 55 and 41 Hv, respectively.

Morisada *et al.,(2009)* studied the effect of Nanostructured tool steel fabricated by combination of laser melting and friction stir processing. A commercially available plate of tool steel (SKD11) was used in this study. The surface of the plate was melted by multi-pass laser heating (1kW, LASERLINE LDF-1000 750) to produce a rapidly solidified zone for the FSP. The FSP was completely carried out in the rapidly solidified zone of the laser treated SKD11. The FSP tool made of hard metal (tungsten carbide based alloy) had a columnar shape (⌀12mm) with a probe (⌀4mm, length: 0.5 mm). No detectable chips and cracks were found on the surface of the tool after the FSP by optical microscopy observation. A constant tool rotating rate of 400rpmwas adopted and the constant travel speed was 400 mm/min. The matrix grains and carbide particles of the SKD11were significantly refined by the laser melting and the FSP. The microstructure and microhardness were evaluated by observations of the grain size and phase of the matrix, and the size and dispersion of the carbide particles. The nanometer-sized microstructure consists of a fine carbide (particle size: ~100 nm) and matrix (grain size~200 nm) as fabricated by the combination of laser melting and the FSP. The carbide is theM7C3 type for the FSPed SKD11with and without laser melting. The carbide formed by laser melting includes a high amount of iron and molybdenum when compared to that formed by the FSP without laser melting. The microstructural constituent of the laser treated SKD11 with and without the FSP is martensite and the retained austenite. The nanostructured SKD11 has an extremely high hardness of about 900HV.

Puviyarasan *et al.,(2011*) Fabricated and Analysed Bulk SiCp Reinforced Aluminium Metal Matrix Composites using Friction Stir Process. The base metal used in the experiment was commercially available AA6063-T4 aluminium alloy rolled plate of a nominal composition and the reinforced particles used was green Sic powder having an average diameter ~3micron and purity ~99.9%.The aluminium plate were cut in a rectangular shape of dimension (100*50*10mm).FSP tool was made of high speed steel and hardened to 60 hrc. The tool pin was 6 mm diameter, shoulder diameter of 18 mm. The tool was rotated in clock wise direction; the tool rotation was kept constant at 1000 rpm. The advancing speed was varied to 30, 40 and 50 mm/min. A groove was cut using a slitting saw cutter of width 1.2, 1.5, 1.8 mm and depth 6 mm exactly in the centre of the specimen plate's .SiC powder was deposited in the groove. After the fabrication of Al MMCs the microstructure of stir zone was observed by optical microscope. The black color spot indicate the presence of SiC. The yellow color indicates the Al base metal. The micro hardness of the composite was tested using 0.5 kg load. The hardness value has increased from 40 Hv to 62 Hv. The micro hardness is also found to be 30% higher than that of the base metal.

Qing Su *et al.,(2005)* studied the microstructure evolution during FSP of high strength aluminium alloys. The base material selected for this investigation was 7075 Al plate of 6mm thickness. A single pass friction stir processed zone was produced at a rotational speed of 350 rpm and a translational speed of 12 cm/min. Microstructures were investigated by TEM. This microstructure evolution suggests that the final microstructures of processed material are strongly dependant on the tool design, processing parameters and cooling rate.

Ramesh et al.,(2012) investigated the effect of multipass FSP on mechanical properties of aluminium alloy 5086. In this process 6mm thick plates were used with dimensions 150X110mm. A hot die steel toll with flat shoulder of 24 mm dia, a cylindrical pin of 6mm dia and 3mm length is used. The separation between the subsequent pass is 3mm. a total of 12 passes were carried out. The machining parameters were kept 1025rpm (fixed) and a variable feed rate of 30mm/min, 50mm/min, 80mm/min, 110mm/min, and 150mm/min. A specially designed fixture was used to hold the plate firmly and a mild steel plate was used as backing. Two types of processing methods were adopted in this study, multipass in which material was allowed to cool back down to room temperature and after that second pass was employed and second method is all subsequent 12 passes were employed continuously without alloying the material to cool down. Further microstructure were checked, for both methods and tensile test and microhardness test was carried out. The result showed that intermittent multipass showed better mechanical properties than continuously multipass process.

Robson et al.,(2010) investigated the mechanism of microstructural evolution during friction stir processing (FSP) of cast magnesium alloy AZ91. Rectangular plates of AZ91 with dimensions 630×115×45mm were industrially ingot cast, following which the surfaces were machined off to give a final plate thickness of 39mm. The machining parameters he selected were 1000rpm speed and 160mm/min advancing speed. Microstructural evolution during friction stir processing of cast magnesium alloy AZ91 has been investigated by means of a breaking–pin study, where the tool pin is induced to break during processing followed by rapid quenching of the surroundings to capture the microstructure.

Sato et al.,(2005) studied the effect of FSP multipass on highly formable Mg alloy plate. In the present study plates of a die cast AZ91D Mg alloy (9 wt% Al and 1 wt% Zn), 2 mm in thickness were used. Multi-pass FSP was applied to this alloy using a general FSP tool the parameters used were a travel speed of 12.0 mm/s and a rotational speed of 800 rpm, and a travel speed of 1.5 mm/s and a rotational speed of 1200 rpm. The microstructure of base metal and FSPed shows the base material contained a lot of porosity which is usually inevitable in die cast alloys, while the FSPed zone does not have any porosity or defects. The grain sizes lie between 2.4 and 2.8 μm in the cold FSPed alloy, while they range between 6.9 and 7.3 μm in the hot FSPed alloy. Average grain sizes were 2.7 and 7.0 μm in the cold and hot FSPed alloys, respectively. The difference in grain size between the cold and hot FSPed alloys can be explained by the heat input during FSP, because the cold FSPed alloy experienced a lower heat input than the hot FSPed alloy during FSP. It should be noted that the grain size profiles are roughly constant throughout the multi-pass FSPed alloys. This result suggests that the multi-pass effect on the grain size is negligible.

Sun *et al.*,(2012) studied the effect of friction stir processing on microstructure and mechanical property of nano-SiCp reinforced high strength Mg bulk composites. Commercial SiC particles and as-casted AZ63 alloy plate (160 mm × 65 mm × 4 mm), were chosen as reinforced particles and base metal separately. A groove with dimension of 2 mm × 160 mm × 2 mm was cut by a milling machine before FSP. The groove was cut along the longitudinal direction of the base metal plate for filling the SiC powder. After filling the powder, another plate in same size with no groove was put upon the first plate, and then the two plates were put upside down. The pin of FSP tool was 4.2 mm in length and 6 mm in diameter, and the shoulder was 20 mm in diameter. The traveling speed of the FSP tool was 20 mm/min and the rotation speed was 1500 rpm. 5 FSP passes were applied in the same position in this study in order to make the reinforced particles distributed uniformly in the nugget zone. Plates without groove were also friction stir processed for 5 passes in the same position as comparison group. The macrostructure of the nugget zone was more uniform. Thus the nugget zone in the transection of FSP materials could be easily identified. The area of the nugget zone was about 9 mm × 4 mm, with no defects observed. With the addition of SiC nanoparticles, the shape of composites nugget zone changed little from the comparison group. Some onion rings and a few small black particles could be observed in the central of the composite. The average Vicker hardness of the base metal, comparison group and composite were 80 Hv, 85 Hv and 109 Hv respectively. The ultimate tensile strength of the composite reached 312 MPa, compared with 160 MPa of the as-casted Mg alloy, 263 MPa of the comparison group. The SiC particles were found both inside the grain and at the grain boundary. The particles stopped dislocation slipping, which improved the mechanical properties of the composite significantly.

Surekha *et al.*,(2009) used multipass FSP to check the corrosion behaviour and microstructural characterization on aluminium AA2219. The traverse speed did not have any significant influence on the corrosion behaviour. Grain refinement was noticed in the alloy after FSP in comparison to base metal. The BM shows a large average grain size of 67.4 μm while the friction stir processed sample showed an average grain size of 6.2 μm in the first pass itself. With subsequent passes the average grain size showed a marginal increase with MS2 and MS3 samples showing 6.7 μm and 7 μm, respectively. On multipassing nugget zone showed lower hardness than base metal. Salt spray tests were carried out in 5% NaCl solution for 100 h to assess the uniform corrosion resistance which showed significantly increase in corrosion resistance as the number of passes increased. The resistance increases as the passes increases as compared to the base metal. The corrosion rate in the base metal is 15.7 mpy where as it is 4.5, 4.0 and 3.8 mpy in MS1, MS2 and MS3 samples.

XingHao *et al.,(2009)* used two pass FSP to produce nanocrystelline microstructure in AZ61 magnesium alloy. TEM images of the microstructure for the one-pass FSP AZ61 alloy, which shows that the microstructure can be only refined to submicron scale, with the grain size of about 500 nm. The two-pass friction stir processing combined with rapid heat sink produces nano-grained structure for AZ61 alloy. The mean grain sizes can be refined to less than 100 nm. The highest microhardness reaches 155 *Hv*, which is almost three times as high as that of the AZ61 substrate.

Yazdipoura *et al.,(2009)* studied the effect of cooling rate on Al5083 subjected to friction stir processing. AFM analysis was also used for studying the surface microstructure of samples. The final structure is considerably refined resulting in the formation of ultrafine grains (UFG) and nanograins ranging from 100 to 500 nm.grains. Thus, the final microstructure consists of submicron grains formed during FSP by dynamic recovery and continuous dynamic recrystallization as well as nanograins formed during meta-dynamic recovery described above. In other words, in the case of latter, the statically formed nanosize nuclei are subjected to rapid cooling and had no chance of growing during the time left the tool and subjected to rapid cooling. The results also show that while heat input to the stir zone may control the 'nuclei size' and 'nucleation' mechanisms, cooling rate may considerably affect the 'grain growth' step.

Zahmatkesh *et al.,(2010)* studied about the microstructural evolution during FSP and their effects on hardness and wear resistance of Al2024 alloy were investigsted. Very fine equiaxed grains, due to the dynamic recrystallization process acting during FSP, are observed in the nugget zone. According to XRD patterns the structure of as-received sample includes CuAl2 and CuMgAl2 intermetallic compounds. Dissolution of CuMgAl2 particles in NZ occurred as a result of an increase in temperature. The NZ exhibited homogeneous and fine equiaxed grains with average size of 4 μm. The maximum hardness was achieved in NZ (110 Hv). FSP was found to be beneficial in improving wear resistance. The high wear behaviour in the NZ is attributed to a lower coefficient of friction and the improved micro-hardness in this region.

2.2 Summary of related literature

Table 2.1 shows brief Summary of related literature

S. No.	Author	Title	Objective	Limitations
1	Elangovan et al., 2008	Influence of tool pin profile and axial force	To optimize the Tensile strength, using different tool pin profiles	Joints fabricated by straight cylindrical pin profiled tool exhibited inferior tensile properties compared to their counterparts
2	Darras et al., 2007	FSP of Magnesium alloy AZ31	To study the microstructure and hardness value	Maximum hardness occurred in the stir zone, increase in r.p.m. decreased the hardness values
3	Surekha et al., 2009	Used multipass FSP process on AA2219	FSP factors influencing Corrosion behaviour	Only the rotation speed was found to influence the corrosion behaviour, while the traverse speed had negligible influence
4	Zahmatkesh et al., 2010	Used FSP on Al2024 alloy	To study the Wear behaviour	FSP was found to be beneficial in improving the wear resistance The maximum hardness was achieved in NZ (110 HV)

36

| 5 | Cavalier et al., 2005 | Used FSP on Aluminium alloy 7075 | Hardness profile of the stirred zone, Tensile tests at room temperature and at higher temperatures | Maximum hardness occurred at a distance of 3.5 mm from the center rather than at the centre. |

2.3 Research Gaps

1. Tool multipass FSP not performed on AL6063 with constant parameters.
2. Microstructure characterization of aluminium alloy after tool multipass have not studied.
3. No impact test had been done on AL6063 after FSP.
4. Microhardness didn't studied after multipass of tool

2.4 Objective of Study

1. To study the effect of tool multipass on AL6063 using FSP.
2. To study the microstructural changes in AL6063.
3. To study impact strength after conducting experiment of multipass on FSP.
4. To study the microhardness after doing FSP on AL6063

2.5 Methodology

- Extrusion of AL6063 in the required form i.e., length, breadth and thickness of the required sample, required composition of different materials in aluminium base.
- Cutting the plate in the required size, and performing the experiment on vertical milling machine with the selected parameters and performing the different number of FSP passes.
- Evaluation of microstructure obtained after performing the experiment on the aluminium samples, and comparing the results of different samples under optical microstructure.
- Characterizing of microhardness of the different samples after conduction of experiment using vicker's hardness tester, and checking the effect of multipass on the sample.
- The impact strength of the different samples is to be studied out to get the optimum results of impact strength of the samples obtained after experimentation.

CHAPTER 3
EXPERIMENTATION

Friction stir processing (FSP) was carried out using a CNC vertical milling machine set up fitted with a specially designed cylindrical FSP tool which was rotated. The base metal (i.e. Al 6063) was fixed in a specially designed fixture that was held on the bed of milling machine. The FSP tool was rotated on the surface of base material to fabricate or machine the surface of Al6063. The surface prepared was characterized for optical microscopy, microhardness and wear test. The results were also obtained for base metal (6063Al) for comparison. The details of various equipments, techniques and procedures used in this study are presented in this chapter.

3.1 Sample Preparation

The sample to be used is of aluminium alloy with dimensions Length 150 mm, Breadth 100 mm, and thickness of 6mm. The alloy selected for the experiment is AL6063. It is an aluminium alloy with magnesium and silicon as the alloying element.

Table 3.1: Showing Physical composition of al 6063 alloy

| Table 3.1: Chemical composition of Al-6063base plate ||
| (Plate dimensions: Thickness: 6mm, Length: 150mm, Width: 100mm) ||
ELEMENT	Wt. (%)
Si	.600
Mg	.619
Cu	.082
Fe	.350
Mn	.044
Zn	.061
Cr	.007
Ti	.015
Ni	.006
Sn	.024
Pb	.046
Al	97.94

It is widely used for architectural fabrication, window and door frames, pipe and tubing, and aluminium furniture. The composition of alloy is:

3.1.1 Application of AL6063

Aluminium alloy 6063 is typically used in:

· Architectural applications

· Extrusions

· Window frames

· Doors

· Shop fittings

· Irrigation tubing

3.1.2 Fabrication of AL6063

Process Rating

- Workability - Cold Average
- Machinability- Average
- Weldability – Gas Excellent
- Weldability – Arc Excellent
- Weldability – Resistance Excellent
- Brazability -Excellent

3.1.3 Supplied Forms

- Square Box Section
- Rectangular Box Section
- Channel
- Equal Angle
- Unequal Angle
- Flat Bar
- Tube

Fig 3.1 Aluminium Alloy AL 6063 Sample

3.1.4 Properties of Aluminium AL6063

The most common tempers for 6063 aluminium are:

• Annealed wrought alloy

• T4 – Solution heat treated and naturally aged

• T6 – Solution heat treated and artificially aged

Table 3.2: Showing Tempers of al 6063 alloy

Temper	**O**	**T4**	**T6**
Minimum Proof Stress 0.2% (MPa)	50	65	160
Minimum Tensile Strength (MPa)	100	130	195
Shear Strength (MPa)	70	110	150
Elongation A5 (%)	27	21	14
Hardness Vickers (HV)	25	50	80

Table 3.3: Showing Physical properties of al 6063 alloy

Property	**Value**
Density	2.70 g/cm3
Melting Point	600°C
Modulus of Elasticity	69.5 GPa
Electrical Resistivity	0.035x10-6 O.m
Thermal conductivity	200 W/m.K
Thermal Expansion	23.5 x 10-6 /K

3.1.5 *Welding of 6063 Aluminium*

- 6063 is suitable for all conventional welding methods.
- Welding wire generally should be alloy 5183 or alloy 4043.
- When maximum electrical conductivity is required use alloy 4043.
- For strength and conductivity use alloy 5346 and increase the size of the weld to compensate for the lower conductivity.

Aluminium alloy 6063A is a variation of 6063 aluminium with greater strength but retains the same good surface finish qualities and affinity for anodising.

3.2 FSP Setup

3.2.1 *Tool Geometry*

The Tool used in FSP is a cylindrical tool made of High Speed Steel. With the total length of 64mm, in which length of shoulder is 60mm and length of pin is 3.5mm, dia of shoulder is 12 mm and dia of pin is 4mm. After cutting the tool in required shape it is heat treated up to M2, to achieve the required hardness of 61 – 62(Rockwell Hardness).

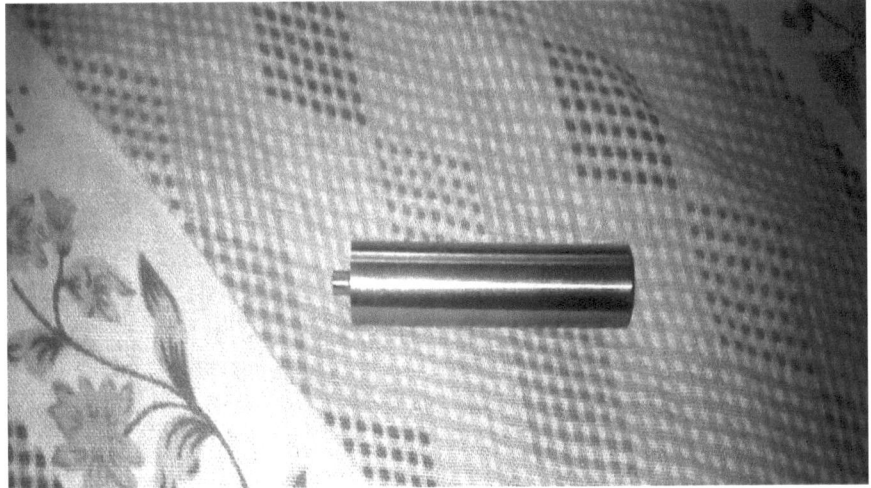

Fig 3.2 FSP Tool

3.2.2 Tool Specifications

Table 3.4: Showing specification of tool

Pin and Shoulder Material	High Speed Steel (Heat Treated M2)
Shoulder Dimension	Dia. =12mm, Length = 60 mm
Pin Dimension	Dia. = 4mm, Length = 3.5mm
Tool Hardness	61-62 (Rockwell Hardness)

3.2.3 CNC vertical milling machine

A CNC Vertical Milling Machine was used for the purpose of carrying out FSP (Fig..3.3). A specially designed FSP tool was clamped in the vertical spindle and the plate shape specimen was clamped on the bed with the help of specially a designed fixture. The specification of the America's Haas automation Inc made CNC are given below in the table.

Fig 3.3: CNC vertical milling machine

Table 3.5: Showing specification CNC Machine (HAAS Automation)

	VF-2SS	
TRAVELS	**S.A.E.**	**METRIC**
X Axis	30"	762 mm
Y Axis	16"	406 mm
Z Axis	20"	508 mm
Spindle Nose to Table (~ max)	24"	610 mm
Spindle Nose to Table (~ min)	4"	102 mm
TABLE	**S.A.E.**	**METRIC**
Length	36 "	914 mm
Width	14 "	356 mm
T-Slot Width	5/8 "	16 mm
T-Slot Center Distance	4.92 "	125.0 mm
Number of Std T-Slots	3	3
Max Weight on Table (evenly distributed)	1500 lb	680 kg
SPINDLE	**S.A.E.**	**METRIC**
Max Rating	30 hp	22.4 kW
Max Speed	12000 rpm	12000 rpm
Max Torque	90 ft-lb @ 2000 rpm	122 Nm @ 2000 rpm
Drive System	Inline Direct-Drive	Inline Direct-Drive
Taper	CT or BT 40	CT or BT 40
Bearing Lubrication	Air/Oil Injection	Air/Oil Injection
Cooling	Liquid Cooled	Liquid Cooled
FEEDRATES	**S.A.E.**	**METRIC**
Rapids on X	1400 in/min	35.6 m/min
Rapids on Y	1400 in/min	35.6 m/min
Rapids on Z	1400 in/min	35.6 m/min
Max Cutting	833 in/min	21.2 m/min
AXIS MOTORS	**S.A.E.**	**METRIC**
Max Thrust X	1995 lb	8874 N
Max Thrust Y	1995 lb	8874 N
Max Thrust Z	3085 lb	13723 N
TOOL CHANGER	**S.A.E.**	**METRIC**
Type	SMTC	SMTC
Capacity	24+1	24+1
Max Tool Diameter (adjacent empty)	5 "	127 mm
Max Tool Diameter (full)	3 "	76 mm
Max Tool Length (from gage line)	11 "	279 mm
Max Tool Weight	12 lb	5 Kg
Tool-to-Tool (avg)	1.6 sec	1.6 sec

Chip-to-Chip (avg)	2.2 sec	2.2 sec
GENERAL	**S.A.E.**	**METRIC**
Air Required	4 scfm, 100 psi	113 L/min, 6.9 bar
Coolant Capacity	55 gal	208 L

3.2.4 Fixture

The fixture for holding the base plate while carrying out FSP was designed in house and fabricated at Dhiman Industries, Bathinda. The fixture consisted of a rectangular base of dimensions 400mm x 200mm x 20mm. Three numbers of square rods having X-section (25mm x25mm) and length 300mm were machined to an accuracy of 5µm. Out of these square rods, two rods were drilled with counter sunk holes (4 Nos.) to adjust allen bolts of size M10. These two rods were fixed at the ends of rectangular base plate. The Sq. rod 2 was also consisted of 3-addition drilled holed consisting of M12 internal threads to accommodate hexagonal bolts of size M12. The third square rod was placed between Sq. rods 1 and 2 and was movable with the help of hexagonal bolts tightened to square rod 2. The base plate to be friction stir process was held tight between square rods 1 and 3. The base plate was also tightened to the rectangular base plate with the help of 2 MS strips (S1 and S2) each screwed to the rectangular base plate with the help of hexagonal bolts (M8). The whole of the fixture shown in (Fig.. 3.4) was fixed with the work plate.

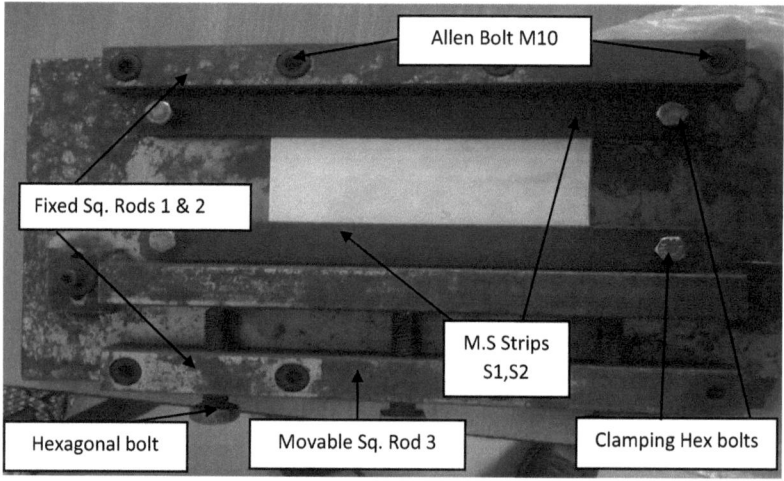

Fig 3.4: Fixture used for holding the specimen during FSP

Fig 3.5 showing the used tool

3.2.5 Machining Parameter

The machine used for experimentation is CNC milling Machine with the constant speed rate of 1100 RPM and Feed Rate of 15mm/min.

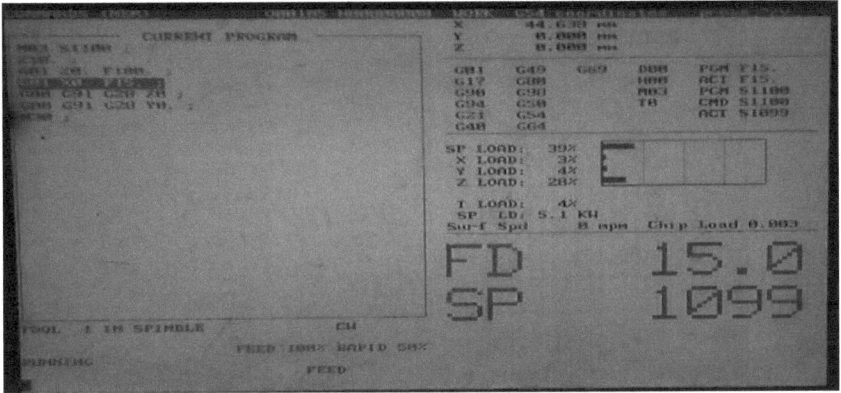

Fig 3.6 showing the experimental parameters

3.2.6 Different Results after experimentation

All the 3 samples are passed under the vertical CNC milling machine (Haas automations) using the cylindrical FSP tool, with the same parameters, the feed rate was kept 15mm/min and speed rate was kept 1100rpm, when first sample is passed under the tool, the surface finish generated not so good. When second sample with two pass is machined the surface finish is quite good as compared to the sample 1, similarly third sample shows very little improvement in surface finish as compared to ms1, ms2. The tool was passed throughout the workpiece transversely forming the tunnelling effect on the workpiece.

All four samples after experiment are shown below.

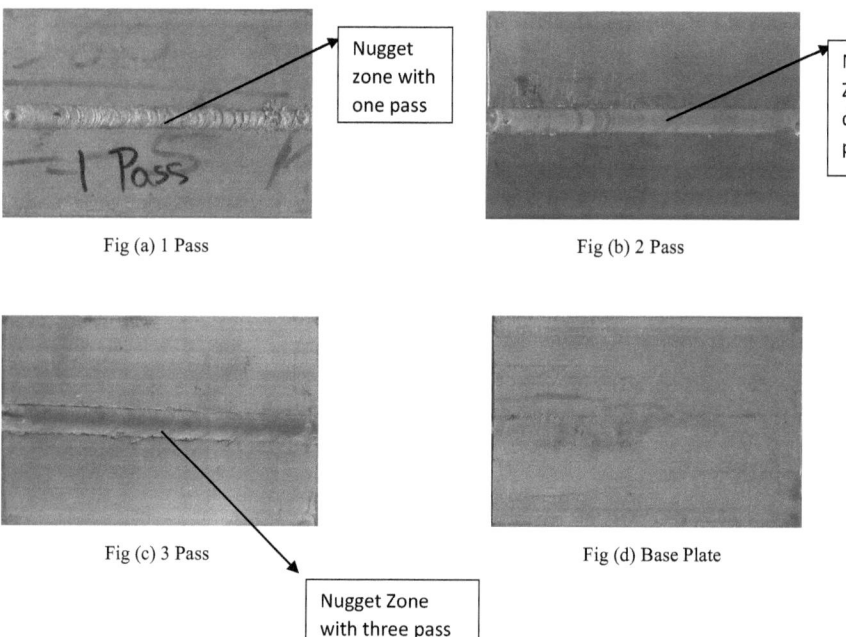

Fig (a) 1 Pass

Fig (b) 2 Pass

Fig (c) 3 Pass

Fig (d) Base Plate

Fig 3.7 shows the sample results obtained after conducting experimental, it is clearly differentiable between the surface finish by seeing the above pictures of different samples.

3.3 Characterization and testing

In order to investigate the effect of surface modification of 60633Al, the FSP samples were subjected to microhardness and wear tests. The microstructures of the samples were observed through optical microscopy. The details of testing and characterization carried out in the study are described in the following section.

3.3.1 Microstructural examination

The microstructure of the samples were observed under inverted optical microscope (Metallurgical microscope with image analyzer, Make-Qualitech) having PC interface and Measurement software as shown in Fig.ure 3.7.

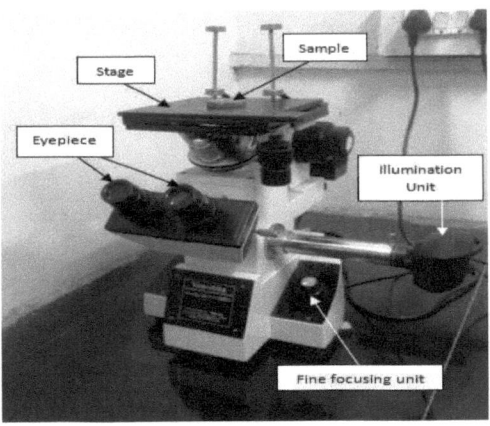

Fig 3.8: Inverted Optical Microscope (Qualitech)

The specimen prepared for viewing their microstructure. Firstly the friction stir processed samples were cut into small sizes along the longitudinal as well as transverse sections. The cutting was performed with the help of hand hacksaw so as to avoid heating of the specimens, which might have cause microstructural changes. The cut pieces were mounted inside thermosetting plastic. The samples were placed inside a hollow pipe (dia.=50mm, length=20mm) with pressed surface downward. A freshly prepared mixture of selfcure polymer powder and selfcure monomer liquid was poured inside the hollow pipe from the top and the mixture was allowed to set for 5 min. The mounted process was carried out on metal base with greased surface so as to avoid adhesion of mounted samples with the metal base.

Fig 3.9: Mounted specimens

The mounted samples (Fig 3.8) were first ground on the flat belt grinder followed by manual grinding on successive grades of emery papers starting from 250 grits to 400, 800 coarse to fine grits of 1600 and 2000. The polishing was done until mirror finished and crack free surface was obtained. The final polishing was done on buffing machine by using diamond paste and continuously running water (Fig 4.9).

Fig 3.10: Buffing Machine

In order reveal the microstructure of the polished specimens, the samples were etched with 5% Nital (0.5ml HF in 99.5ml H_2O) for 5 to 10 sec followed by washing with wet cotton. The etched surface was dried with forced hot air to remove traces of moisture, if any.

3.3.2 Microhardness measurement

Microhardness of the friction stir processed samples were measured at various at nugget zone with Microvicker hardness tester, Made Akashi (Model MVK-H2).The size of indention was measured with the help of micrometer and eyepiece (having magnification of the order 10X and 40X) fitted on the hardness tester. The impression was created with the help of indentor (Pyramid shape) by applying load of 100gms, for a dwell time of around 10 sec. The specimen table was provided with micrometer and rack and pinion system to facilitate movements of the specimen along both X and Y axes. Where HV is microhardness of the samples, P is applied load in Kg and d is size (diagonal) of the impression in mm.

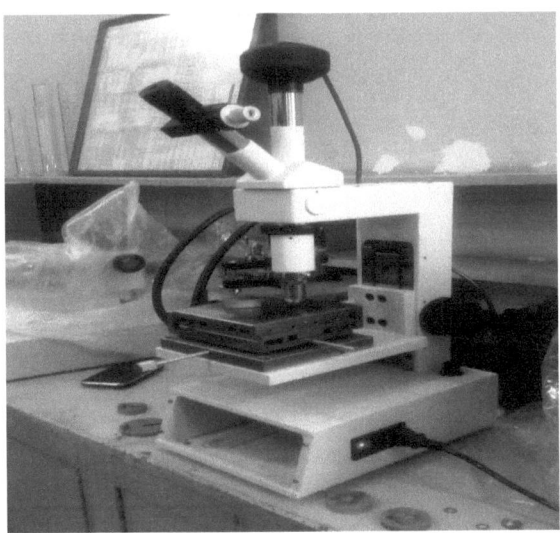

Fig 3.11: Micro hardness tester (Akashi made)

3.3.3 Izod Impact test

Izod impact testing is an ASTM standard method of determining the impact resistance of materials. An arm held at a specific height (constant potential energy) is released. The arm hits the sample and breaks it. From the energy absorbed by the sample, its impact energy is determined. A notched sample is generally used to determine impact energy and notch sensitivity. . The test is similar to the Charpy impact test but uses a different arrangement of the specimen under test. The Izod impact test differs from the Charpy impact test in that the sample is held in a cantilevered beam configuration as opposed to a three-point bending configuration.

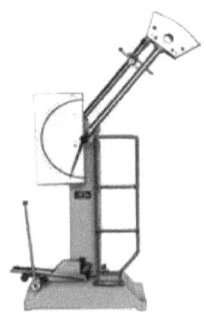

Fig 3.12 Impact testing machine

All the three samples were tested using izod test. The samples were cut into 75 x 10x 6. And the strength of samples were measured in joules. The sample proned to impact test was not broken from notch due to the ductility of aluminium that is high. But the pictures are shown below which depicts the effect of impact on the aluminium samples. The sample 1 which was single passed FSP sample have got broken or tear off from the notch with respect to sample 2 and 3(2pass and 3 pass) which only showed bend in the sample rather than breaking ore tearing off.

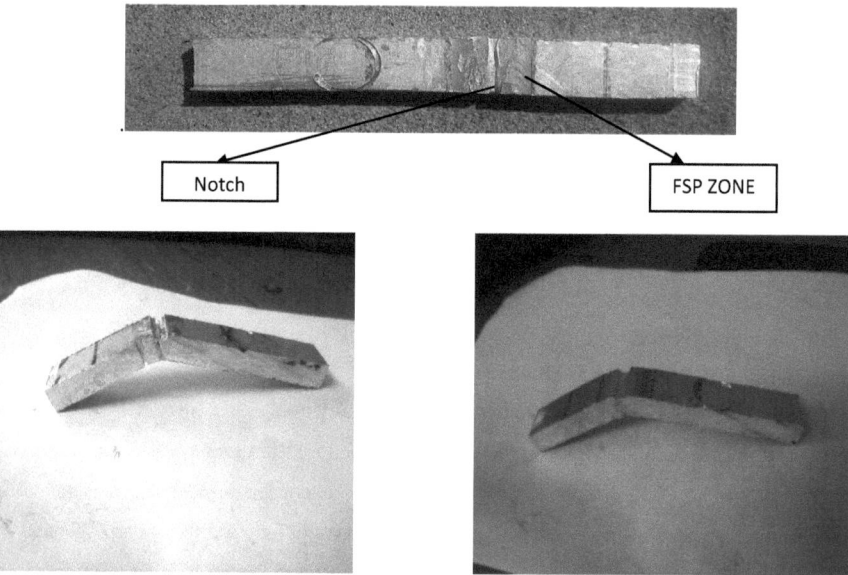

Fig (a) shows the effect of impact on sample 1   Fig (a) shows the effect of impact on sample 2

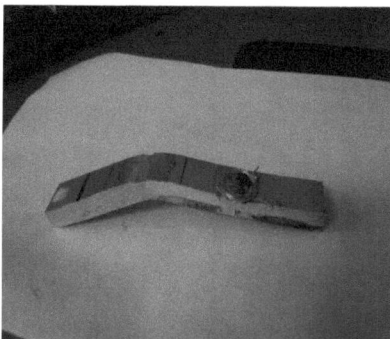

Fig (c) shows the effect of impact on sample 3

Fig 3.13 shows different results after impact test was carried out on different samples

3.3.4 Rockwell hardness test

The Rockwell hardness test was done on the Fsped sample, using scale B (generally for aluminium) with a load of 100kgs, with the indenter ball size 1/16" ball. The below is the table showing the different load used for different material and the indenter used

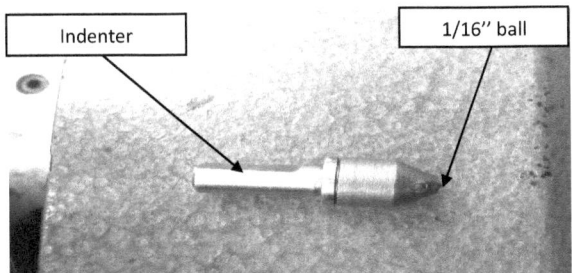

Fig 3.14 shows the type of indenter used to perform Rockwell test

Fig 3.15 shows the Rockwell hardness tester holding the sample

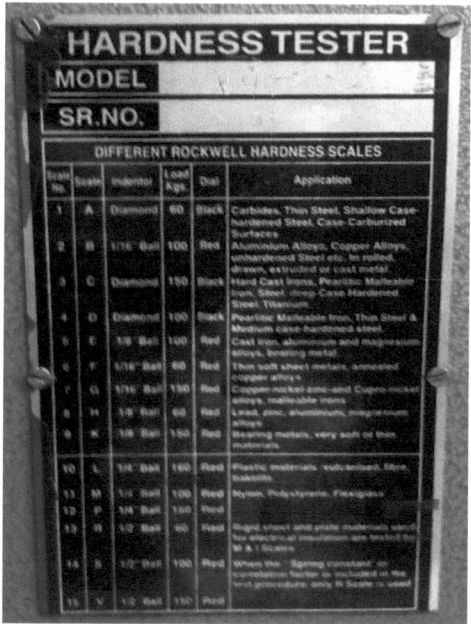

Fig 3.16 showing the hardness tester specifications

The test is performed using indentation tool, at different areas of the workpiece, by this hardness can be measured at different points, and after taking 3 readings at each sample the average of the reading is calculated and stated as Rockwell hadness in HRB scale used for aluminium alloys

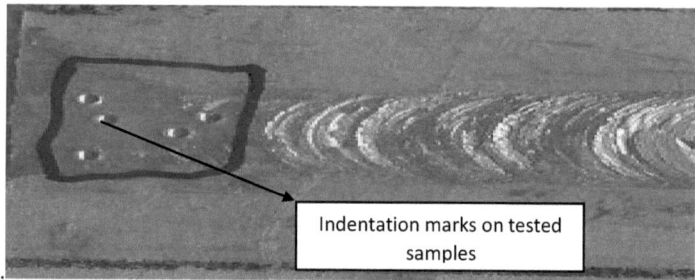

Fig 3.17 showing the hardness tested sample

CHAPTER 4

RESULTS AND DISCUSSION

Al-Alloy (6063Al) based surface, was successfully processed by friction stir processing (FSP) on CNC milling machine. The spindle rpm and feed rate were kept respectively as 1100 and 15mm/min during the course of FSP. The FSP tool used for the purpose is already described on chapter 3.

4.1 Evaluation of mechanical properties

Subsequent to friction stir processing, the samples were cut to evaluate the mechanical properties like microhardness, microstructure, impact strength of the FSP alloy and base alloy.

4.1.1 Microhardness evaluation

Microhardness of the samples was measured between the nugget zones. Microhardness profiles measured across the top surface of the sample are shown in Fig 4.1. The hardness measurements were performed on top side of FSP zone. Similarly the microhardness of the base alloy was also measured. It is observed that value of hardness in the FSPed zone is significantly higher than that observed for the base alloy. The hardness profile shows that the hardness is maximum (62HV) of FSPed zone MS 3(3 Pass). The microhardness measurements of 2nd pass was observed MS 2 50 HV and of first pass was 60 HV. However, the microhardness of the base alloy remains almost uniform, ranging between 47 to 49HV. Where as in second pass of FSPed composite, the microhardness was lower compared to first pass and third pass Al alloy this is due to the material become more softening and annealing effect of heat input during the process. The FSP with the SiC particles is considered to make fine grains more effectively due to the enhancement of the induced strain and the pinning effect the SiC particles. Microhardness of the FSPed samples was also measured across the transverse section at different locations. The subsurface hardness is also maximum in the nugget/stir zone and decreases significantly on moving away from the centre of FSPed zone. The subsurface hardness, at a given depth, is significantly higher than the hardness measured at the corresponding location on the top surface.

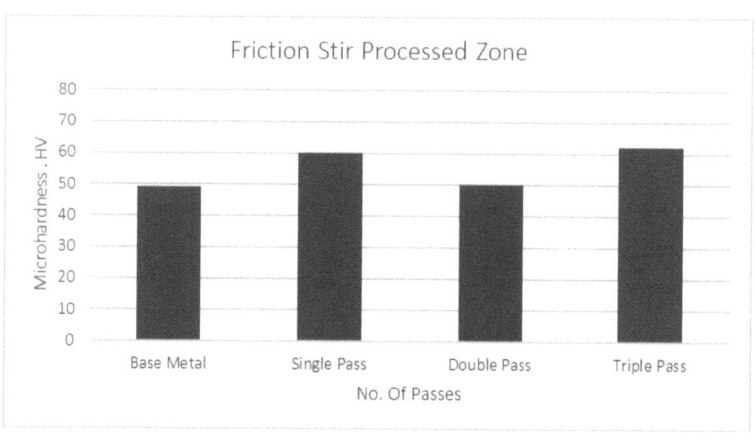

Fig 4.1 Graph Showing the comparison between Microhardness results

This result is expected since the temperature reached during processing increases with the rotational speed which leads to more softening due to grain growth, according to the well-known Hall–Petch relationship which states that the hardness is inversely proportional to grain size. Therefore, more refinement can be achieved at lower rotational speeds. As the temperature decreases from the surface towards the bottom of the sheet, the hardness increases. Higher temperature leads to more softening and more grain growth. The hardness profile shows that the maximum hardness occurs at the center of the deformation zone (nugget). An interesting result is also illustrated the minimum hardness values are observed at the interface between the thermomechanical affected zone (TMAZ) and the heat affected zone (HAZ), where the hardness values are smaller than those of the base material. The hardness drop at the heat affected zone is due to the fact that there is no mechanical deformation (stirring) at that zone; however the peak temperature reached is enough to soften the material near the nugget.

4.1.2 Rockwell hardness test

The Rockwell hardness test was conducted using a indenter with ball of 1/16'' and was forced into the surface of test sample in two operations and the permanent increase in depth of identification under specified conditions was measured. The depth of indentation is direct measure of Rockwell hardnesss. The hardness was tested using all the samples at different FSPed zone. There were 2-3 readings were taken and then average reading is calculated. In this test the third pass sample showed little improve in hardness with respect to base metal. The Rockwell hardness received for base metal was 24 HRB. The first pass shows slight increase in hardness upto 26 HRB and in second pass the hardness decreased to 25 HRB and after third pass the hardness increased to 28 HRB. The above calculated reading was the average to 3 readings taken from different points on the sample and hardness was calculated using the formula:

$$\text{Mean value R} = \frac{R_1 + R_2 + R_3}{3}$$

Below the graph comparing different Rockwell hardness result. The Rockwell hardness at the centre of FSPed zone is maximum

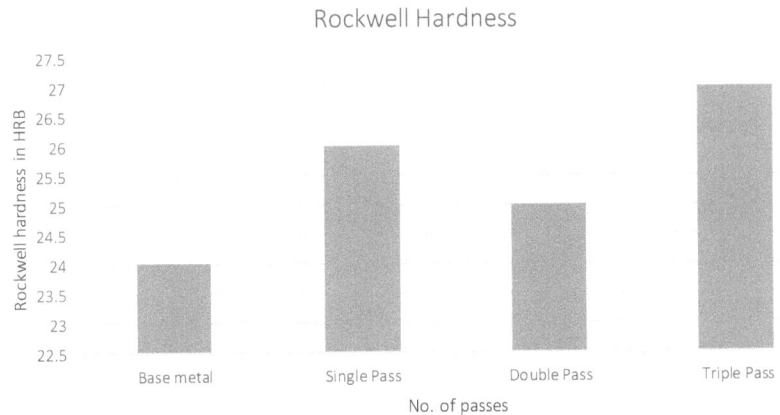

Fig 4.3 shows comparison between Rockwell hardness

The change in Rockwell hardness as compared to the base metal is due to the heat treatment of the FSPed samples because as the toll moves transversely on the workpiece causes friction and due to that friction high heat was generated which effects the grain structure of the samples thus causing the change in grain structure that results in change in the hardness value of the sample.

4.1.3 *Optical microscopy*

Samples for metallographic examination were cut from the surface of base alloy and FSPed sample. These samples were wet grounded using various grades of emery paper and polished to attain mirror finish by using buffing machine and alumina paste. The etching of the samples was done with the reagent 0.5 ml HF (40%) diluted with 99.5 ml distilled H_2O before viewing the microstructure under optical microscope. Optical micrographs of Al-6063 base alloy and FSPed samples were viewed at magnification 100 x and are depicted in Fig 4.4. From these micrographs it is observed that the grain size of FSPed sample is relatively finer than the base alloy. The coarse grains size observed in base alloy leads to high ductility and low hardness than the FSPed sample.

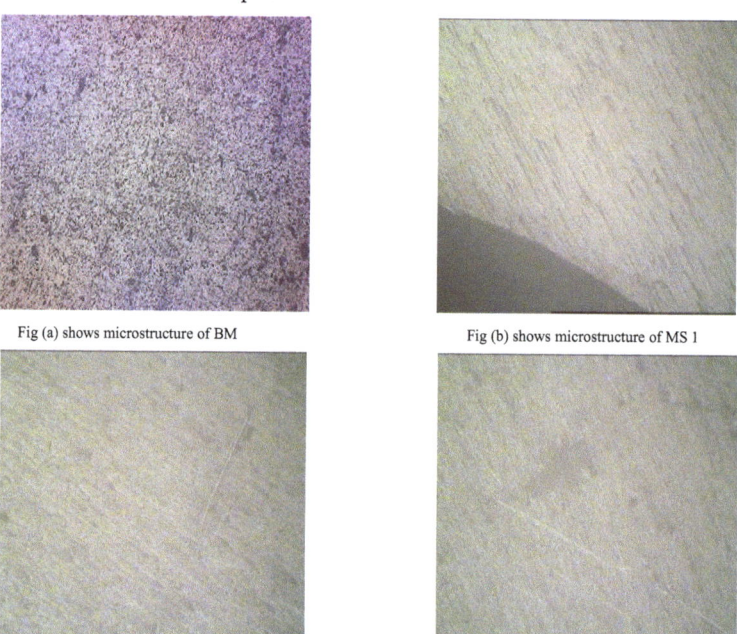

Fig (a) shows microstructure of BM   Fig (b) shows microstructure of MS 1

Fig (c) shows microstructure of MS 2   Fig (d) shows microstructure of MS 3

Above Fig 4.4 shows the microstructure after different passes on aluminium 6063 checked under 100x magnification.

But after the processing of aluminium by FSP, refinement of grain takes place due to high temperature produced by friction between the surfaces of the FSP tool and sample. The fine grain size observed in FSPed sample is responsible for its high hardness than the base alloy. The grain refinement in the matrix may be due to the occurrence of dynamic recrystallization phenomenon as a result of disruptive mechanical action of the tool pin provided in the FSP tool. The base metal had a coarse grain structure, after performing single pass FSP very fine uniformly discontinuity alloy silicides in a matrix of aluminium, some surface defects were observed in the base metal. In second pass very fine distributed alloy silicides in a matrix of aluminium was observed. And in third pass very fine uniformly distributed alloy silicides in a matrix of aluminium, no cavity/ crack/discontinuity observed between processed zone and base metal. The formation of these fine grains during FSP can be attributed to dynamic recrystallization and a low heat input during FSP results in an exceptionally fine grain structure along with dissolution of the precipitates. When the FSP is carried out with higher heat inputs, the grains in the nugget are coarser. On multipassing, the heat input is slightly increased and hence the small increase in grain size on multipassing can be attributed to this factor. Several researchers have suggested that there is a difference in plastic flow behaviour of materials during FSP. It is therefore likely that these microstructural differences resulted from the different flow behaviour and heat input. It was reported that higher rotational speed and lower travelling speed caused more heat input, and the tool supplied shear force to make the silicides particles flow and disperse in wider region. In order to supply enough frictional and shear force for covered particles and avoid them to agglomerate, the shoulder and process parameters play a dominant role.

4.1.4 Izod impact result

The impact test simulates the service conditions often encountered in transportation, agricultural and construction equipment which are frequently subjected to impact and shock loads during sudden blows and stopages. The stress induced during impact loading are many more times the stress induced during gradual loading.

The izod impact test was conducted on all the FSPed sample, using impact testing machine. Which showed decrease in impact strength as compared to the base metal AL6063. It was observed the samples were not totlly broken from the notchj due to the high ductility of aluminium alloy sample. The result obtained after testing the base plate was, the base metal have impact strength of 13 Joule, that subsequent decreased in first pass to 10 Joule, in the second pass the strength increased than 1st pass sample to 11 joule, and in third pass sample it little bit increased to 12 joule. The obtained result showed the impact strength was not improved very much as compared to base metal. The result comparison shown below in the graph.

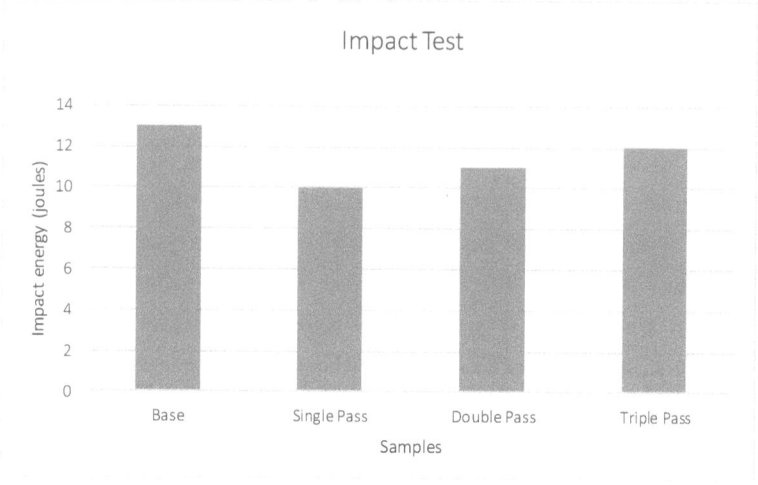

Fig 4.5 shows comparison between impact strength

CHAPTER 5

CONCLUSION AND RECOMMENDATIONS

From the above study it is concluded that the FSP mutipass had effected AL6063, with little increase in hardness that depends upon the microstructure of the samples as the result showed FSP triple pass showed better results than the base metal and first and second pass samples. The microhardness is increased to 62 Hv. The triple pass FSP sample showed very fine uniformly distributed alloy silicide in a matrix of aluminium without any crack with respect to that of single pass, double pass, base metal which also coarse microstructure, and second pass have fine distributed silicides but not uniform. Single pass miocrostructure showed some surface defects with distribution of silicide shows discontinuity. So the microhardness is directly related with the fine microstructure with uniformly distributed silicides in al matrix.

The impact strength does not showed any improvement, the impact resistance of the FSPed sample had been decreased as compared to the base metal.

5.1 Scope for Future Work

- Variability in machining parameters like speed, feed can be changed.
- Different tool profile can be used to perform multipass.
- Al6063 can be embedded with Sic or magnesium particles or powder and Multipass FSP can be performed.
- FSP can be performed after a subsequent pass material should let to cool down at room temperature and another pass can be performed after that.
- Another tests like tensile, Scan electron microscopy, XRD, wear and corrosion tests can be conducted.
- Can be applied on different material like copper, steel, iron.

REFERENCES

- A.Yazdipoura, A. Shafiei Mc, K. Dehghanib; *"Modeling the microstructural evolution and effect of cooling rate on the nanograins formed during the friction stir processing of Al5083"*. Materials Science and Engineering A 527 (2009) 192–197

- A. Devaraju, A. Kumar, A. Kumaraswamy, B. Kotiveerachari *"Influence of reinforcements (SiC and Al2O3) and rotational speed on wear and mechanical properties of aluminium alloy 6061-T6 basedsurface hybrid composites produced via friction stir processing"* Materials and Design 51 (2013) 331–341

- Adem Kurt, Ilyas Uygur, Eren Cete *"Surface modification of aluminium by friction stir processing"* Journal of Materials Processing Technology 211 (2011) 313–317

- B.Zahmatkesh, M.H. Enayati, F. Karimzadeh. *"Tribological and microstructural evaluation of friction stir processed Al2024 alloy"*. Materials and Design 31 (2010) 4891–4896

- B. Zahmatkesh, M.H. Enayati; *"A novel approach for development of surface nanocomposite by friction stir processing"* Materials Science and Engineering (2010)

- B.M. Darras, M.K. Khraisheh. F.K. Abu-Farha, M.A. Omar; *"Friction stir processing of commercial AZ31 magnesium alloy"* .Journal of Materials Processing Technology 191 (2007) 77–81

- C.F. Chen, P.W. Kao, L.W. Chang, and N.J. Ho *"Effect of Processing Parameters on Microstructure and Mechanical Properties of an Al-Al11Ce3-Al2O3 In-Situ Composite Produced by Friction Stir Processing"* DOI: 10.1007/s11661-009-0115-8 The Minerals, Metals & Materials Society and ASM International 2009

- C. I. Chang Y. N. Wang H. R. Pei, C. J. Lee, X. H. Du, J. C. Huang *"Microstructure and Mechanical Properties of Nano-ZrO2 and Nano-SiO2 Particulate Reinforced AZ31-Mg Based Composites Fabricated by Friction Stir Processing"* Key Engineering Materials Vol. 351 (2007) pp. 114-119

- C.I. Chang,a X.H. Dua,b and J.C. Huang; *"Producing nanograined microstructure in Mg–Al–Zn alloy by two-step friction stir processing"* Scripta Materialia 59 (2008) 356–359

- C.J. Hsu, P.W. Kao, N.J. Ho *"Ultrafine-grained Al–Al2Cu composite produced in situ by friction stir processing"* Scripta Materialia 53 (2005) 341–345

- C.J. Hsu, C.Y. Chang, P.W. Kao, N.J. Ho, C.P. Chang; *"Al–Al3-Ti nanocomposites produced in situ by friction stir processing"*. Acta Materialia 54 (2006) 5241–5249

- C.J. Hsu, P.W. Kao, N.J. Ho; *"Intermetallic-reinforced aluminium matrix composites produced in situ by friction stir processing"*. Materials Letters 61 (2007) 1315–1318
- Charit I, Mishra R.S *"High strain rate super plasticity in a commercial 2024 Al alloy via friction stir processing"*. Mater Eng A pp 290–6 (2010)
- Chen Ti-ju, Zhu Zhan-ming, LI Yuan-dong, Ma Ying, Hao Yuan *"Friction stir processing of thixoformed AZ91D magnesium alloy and fabrication of Al-rich surface"* Trans.Non-ferrous Met. Soc. China 20(2010) 34-42
- Devaraju Aruri, Adepu Kumar& B Kotiveerachary *"Effect of III-Pass on Microstructure, Micro Hardness and Static Immersion Corrosion Resistance of AA6061-T6/SiCp Surface Composite Fabricated by Friction Stir Processing"* International Journal of Applied Research In Mechanical Engineering (IJARME), ISSN: 2231–5950, Volume-1, Issue-2, 2011
- Dharmpal Deepak, Ripandeep Singh Sidhu, V.K Gupta; *" Preparation of 5083 Al-SiC surface composite by friction stir processing and its mechanical characterization "* International Journal of Mechanical Engineering ISSN : 2277-7059 Volume 3 Issue 1 (January 2013)
- Don-Hyun CHOI, Yong-Hwan KIM, Byung-Wook AHN, Yong-I KIM, Seung-Boo JUNG; *"Microstructure and mechanical property of A356 based composite by friction stir processing"* Trans. Nonferrous Met. Soc. China 23(2013) 335−340
- Douglas C. Hofmann, Kenneth S. Vecchio; *"Thermal history analysis of friction stir processed and submerged friction stir processed aluminium"*. Materials Science and Engineering A 465 (2007) 165–175
- DU XingHao & WU BaoLin; *"Using two-pass friction stir processing to produce nanocrystalline microstructure in AZ61 magnesium alloy"*. Sci China Ser E-Tech Sci | Jun. 2009 | vol. 52 | no. 6 | 1751-1755
- Essam R.I. Mahmouda, Makoto Takahashi, Toshiya Shibayanagi, Kenji Ikeuchi *"Wear characteristics of surface-hybrid-MMCs layer fabricated on aluminium plate by friction stir processing"* Wear 268 (2010) 1111–1121
- F.C. Liu, B.L. Xiao, K. Wang, Z.Y. Ma *"Investigation of superplasticity in friction stir processed 2219Al alloy"* Materials Science and Engineering A 527 (2010) 4191–4196

- H.R. Akramifard, M. Shamanian, M. Sabbaghian, M. Esmailzadeh; *"Microstructure and mechanical properties of Cu/SiC metal matrix composite fabricated via friction stir processing"* Materials and Design 54 (2014) 838–844.
- Hsiang-Ching Chen *"Effects of Friction Stir Process and Stabilizing Heat Treatment on the Tensile and Punch-Shear Properties of Mg9Li2Al1Zn Magnesium Alloy"* Materials Transactions, Vol. 54, No. 4 (2013) pp. 505 to 511
- I.S. Lee, P.W. Kao, N.J. Ho; *"Microstructure and mechanical properties of Al-Fe in situ nanocomposite produced by friction stir processing"*. Intermetallics 16 (2008) 1104–1108
- Jian-Qing Su, Tracy W. Nelson, Colin J. Sterling *"Microstructure evolution during FSP of high strength aluminium alloys"* Materials Science and Engineering A 405 (2005) 277–286
- K. Elangovan & V. Balasubramanian & M. Valliappan; *"Influences of tool pin profile and axial force on the formation of friction stir processing zone in AA6061 aluminium alloy"*. Int J Adv Manuf Technol (2008) 38:285–295
- K.M.Ramesh, S.Pradeep, Vivek Pancholi *"Multipass Friction Stir Processing and its effect on mechanical properties of aluminium alloy 5086"* The minerals, Metals and materials society and ASM international 2012
- K. Sun, Q.Y. Shi, Y.J. Sun, G.Q. Chen *"Microstructure and mechanical property of nano-SiCp reinforced high strength Mg bulk composites produced by friction stir processing"* Materials Science and Engineering A 547 (2012) 32–37
- K. Surekha, B.S. Murty, K. Prasad Rao; *"Microstructural characterization and corrosion behavior of multipass friction stir processed AA2219 aluminium alloy"*. Solid State Sciences 11 (2009) 907–917
- L. Karthikeyan, V.S. Senthilkumar, K.A. Padmanabhan *"On the role of process variables in the friction stir processing of cast aluminium A319 alloy"* Materials and Design 31 (2010) 761–771
- Liming Ke, Chunping Huanga,b, Li Xinga, Kehui Huanga; *"Al–Ni intermetallic composites produced in situ by friction stir processing"*. Journal of Alloys and Compounds S0925-8388(10)01189-8 2010.
- M.Puviyarasan, C.Praveen *"Fabrication and Analysis of Bulk SiCp Reinforced Aluminium Metal Matrix Composites using Friction Stir Process"* World Academy of Science, Engineering and Technology 58 2011

- Manisha Dixit, Joseph W. Newkirk and Rajiv S. Mishra; *"Properties of friction stir-processed Al 1100–NiTi composite"*. Scripta Materialia 56 (2007) 541–544
- N. Sun and D. Apelian *"Friction Stir Processing of Aluminium Cast Alloys for High Performance Applications"* JOM • November 2011
- P. Cavalierea, A. Squillace; *"High temperature deformation of friction stir processed 7075 aluminium alloy"*. Materials Characterization 55 (2005) 136–142
- P. Cavaliere , P.P. De Marco; *"Superplastic behaviour of friction stir processed AZ91 magnesium alloy produced by high pressure die cast"* Journal of Materials Processing Technology 184 (2007) 77–83
- R.S. Mishra, Z.Y. M *"Friction stir welding and processing"* Materials Science and Engineering R 50 (2005) 1–78
- Ranjit Bauri, Devinder Yadav, G. Suhas *"Effect of friction stir processing (FSP) on microstructure and properties of Al–TiC in situ composite"* Materials Science and Engineering A 528 (2011) 4732–4739
- S.R. Anvari , F.Karimzadeh, M.H.Enayati *"Wear characteristics of Al–Cr–O surface nano-composite layer fabricated on Al 6061 plate by friction stir processing"* Wear 304(2013)144–151
- Scott F *"New friction stir techniques for dissimilar materials processing"* Manufacturing Letters 1 (2013) 21–24
- Srinivasan Swaminathan, Keiichiro oh-ishi, Alexander P.zhilyaev, Christian B. Fuller, Blair London, Murray W.mahoney, Terry R. Mcnelley *"Peak Stir Zone Temperatures during Friction Stir Processing"* The Minerals, Metals & Materials Society and ASM International 2009
- Tanya L.Giles, Keiichiro oh-ishi, Alexander P.zhilyaev, Christian B. Fuller, Blair London, Murray W.mahoney, Terry R. Mcnelley; *"The Effect of Friction Stir Processing on the Microstructure and Mechanical Properties of an Aluminium Lithium Alloy"*. Metallurgical and Materials Transactions a Volume 40A, January 2009
- Terry Khaled *"An outsider look at friction stir welding"* ANM-112N-05-06 (July 2005)
- Y.S.Sato, S.H.C.Park, A.Matsunaga, A,.Honda, H.Kokawa *"Novel production for highly formable Mg alloy plate"* Journal of material science 40 (2005) 637–642

- Y. Morisada, H. Fujii , T. Mizuna, G. Abeb, T. Nagaoka, M. Fukusumi *"Nanostructured tool steel fabricated by combination of laser melting and friction stir processing"* Materials Science and Engineering A 505 (2009) 157–162
- Y. Morisada, H. Fuji, T. Nagoya, M. Fukusumi *"MWCNTs/AZ31 surface composites fabricated by friction stir processing"* Materials Science and Engineering a 419 (2006) 344–348.
- Yong X. Gan, Daniel Solomon and Michael Reinbolt " *Friction Stir Processing of Particle Reinforced Composite Materials* " *Materials* 2010, *3*, 329-350; doi:10.3390/ma3010329
- Yupei Jiang, Xuyue Yang, Hiromi Miura and Taku Sakai *"Nano-SiO2 particles reinforced magnesium alloy produced by friction stir processing"* Rev. Adv. Mater.Sci.33(2013) 29-32
- Z.Y. Ma. *Friction Stir Processing Technology: A Review.* The Minerals, Metals & Materials Society and ASM International 2008
- *http://en.wikipedia.org/wiki/Friction_stir_processing*
- *http://www.esabna.com/us/en/education/knowledge/qa/what-is-friction-stir-welding-of-aluminium.cfm*